A CLIMATE SERVICES VISION

first steps toward the future

Board on Atmospheric Sciences and Climate
Division on Earth and Life Studies
National Research Council

NATIONAL ACADEMY PRESS
WASHINGTON, D.C.

NATIONAL ACADEMY PRESS • 2101 Constitution Ave., NW • Washington, DC 20418

NOTICE: The project that is the subject of this report was approved by the Governing Board of the National Research Council, whose members are drawn from the councils of the National Academy of Sciences, the National Academy of Engineering, and the Institute of Medicine. The members of the committee responsible for the report were chosen for their special competences and with regard for appropriate balance.

Support for this project was provided by the National Science Foundation under Grant No. ATM-9814235, U.S. Environmental Protection Agency, National Oceanic and Atmospheric Administration, and National Aeronautics and Space Administration. Any opinions, findings, and conclusions or recommendations expressed are those of the authors and do not necessarily reflect those of the sponsors or their subagencies.

International Standard Book Number 0-309-08256-0

Additional copies of this report are available from:

National Academy Press
2101 Constitution Avenue, NW
Lockbox 285
Washington, DC 20055
800-624-6242
202-334-3313 (in the Washington metropolitan area)
http://www.nap.edu

Printed in the United States of America

Copyright 2001 by the National Academy of Sciences. All rights reserved.

THE NATIONAL ACADEMIES
National Academy of Sciences
National Academy of Engineering
Institute of Medicine
National Research Council

The **National Academy of Sciences** is a private, nonprofit, self-perpetuating society of distinguished scholars engaged in scientific and engineering research, dedicated to the furtherance of science and technology and to their use for the general welfare. Upon the authority of the charter granted to it by the Congress in 1863, the Academy has a mandate that requires it to advise the federal government on scientific and technical matters. Dr. Bruce M. Alberts is president of the National Academy of Sciences.

The **National Academy of Engineering** was established in 1964, under the charter of the National Academy of Sciences, as a parallel organization of outstanding engineers. It is autonomous in its administration and in the selection of its members, sharing with the National Academy of Sciences the responsibility for advising the federal government. The National Academy of Engineering also sponsors engineering programs aimed at meeting national needs, encourages education and research, and recognizes the superior achievements of engineers. Dr. Wm. A. Wulf is president of the National Academy of Engineering.

The **Institute of Medicine** was established in 1970 by the National Academy of Sciences to secure the services of eminent members of appropriate professions in the examination of policy matters pertaining to the health of the public. The Institute acts under the responsibility given to the National Academy of Sciences by its congressional charter to be an adviser to the federal government and, upon its own initiative, to identify issues of medical care, research, and education. Dr. Kenneth I. Shine is president of the Institute of Medicine.

The **National Research Council** was organized by the National Academy of Sciences in 1916 to associate the broad community of science and technology with the Academy's purposes of furthering knowledge and advising the federal government. Functioning in accordance with general policies determined by the Academy, the Council has become the principal operating agency of both the National Academy of Sciences and the National Academy of Engineering in providing services to the government, the public, and the scientific and engineering communities. The Council is administered jointly by both Academies and the Institute of Medicine. Dr. Bruce M. Alberts and Dr. Wm. A. Wulf are chairman and vice chairman, respectively, of the National Research Council.

BOARD ON ATMOSPHERIC SCIENCES AND CLIMATE

ERIC J. BARRON (*Chair*), Pennsylvania State University, University Park, Pennsylvania
SUSAN K. AVERY, Cooperative Institute for Research in Environmental Sciences, University of Colorado, Boulder, Colorado
RAYMOND J. BAN†, The Weather Channel, Inc., Atlanta, Georgia
HOWARD B. BLUESTEIN, University of Oklahoma, Norman, Oklahoma
STEVEN F. CLIFFORD, National Oceanic and Atmospheric Administration, Boulder, Colorado
GEORGE L. FREDERICK, Vaisala Meteorological Systems, Boulder, Colorado
MARVIN A. GELLER*, State University of New York, Stony Brook, New York
CHARLES E. KOLB*, Aerodyne Research, Inc., Billerica, Massachusetts
JUDITH L. LEAN, Naval Research Laboratory, Washington, DC
MARGARET A. LEMONE†, National Center for Atmospheric Research, Boulder, Colorado
MARIO J. MOLINA†, Massachusetts Institute of Technology, Cambridge, Massachusetts
ROGER A. PIELKE, JR., Cooperative Institute for Research in Environmental Sciences, University of Colorado, Boulder, Colorado
MICHAEL J. PRATHER, University of California, Irvine, California
WILLIAM J. RANDEL†, National Center for Atmospheric Research, Boulder, Colorado
ROBERT T. RYAN, WRC-TV, Washington, DC
MARK R. SCHOEBERL*, NASA Goddard Space Flight Center, Greenbelt, Maryland
JOANNE SIMPSON*, NASA Goddard Space Flight Center, Greenbelt, Maryland
THOMAS F. TASCIONE, Sterling Software, Inc., Bellevue, Nebraska
ROBERT A. WELLER, Woods Hole Oceanographic Institution, Woods Hole, Massachusetts
ERIC F. WOOD, Princeton University, Princeton, New Jersey

Ex Officio Members

DONALD S. BURKE, Johns Hopkins University, Baltimore, Maryland
DARA ENTEKHABI, Massachusetts Institute of Technology, Cambridge, Massachusetts
EUGENE M. RASMUSSON, University of Maryland, College Park, Maryland
EDWARD S. SARACHIK, University of Washington, Seattle, Washington

NRC Staff

ELBERT W. (JOE) FRIDAY, JR., Director
LAURIE S. GELLER, Program Officer
PETER A. SCHULTZ, Senior Program Officer
VAUGHAN C. TUREKIAN, Program Officer
DIANE GUSTAFSON, Administrative Assistant
ROBIN MORRIS, Financial Associate
TENECIA A. BROWN, Project Assistant
CARTER W. FORD, Project Assistant

† Beginning March 2001
* Ending December 2000

ACKNOWLEDGMENTS

This report has been reviewed in draft form by individuals chosen for their diverse perspectives and technical expertise, in accordance with procedures approved by the National Research Council's Report Review Committee. The purpose of this independent review is to provide candid and critical comments that will assist the institution in making its published report as sound as possible and to ensure that the report meets institutional standards for objectivity, evidence, and responsiveness to the study charge. The review comments and draft manuscript remain confidential to protect the integrity of the deliberative process. We wish to thank the following individuals for their review of this report: Mark R. Abbott, Oregon State University; Lee E. Branscome, Environmental Dynamics Research, Inc.; Stanley A. Changnon, Illinois State Water Survey; Richard M. Goody, Harvard University; William E. Gordon, Rice University; Franklin W. Nutter, Reinsurance Association of America; Maria A. Pirone, WSI Corporation; and Jack Williams, *USA Today*.

Although the reviewers listed above have provided constructive comments and suggestions, they were not asked to endorse the conclusions or recommendations, nor did they see the final draft of the report before its release. The review of this report was overseen by S. George Philander (Princeton University) and Louis Lanzerotti (Bell Laboratories). Appointed by the National Research Council, they were responsible for making certain that an independent examination of this report was carried out in accordance with institutional procedures and that all review comments were carefully considered. Responsibility for the final content of this report rests entirely with the authoring committee and the institution.

PREFACE

The uses of climate information are evolving rapidly. In popular usage, climate refers to the average weather or its variations over a period of time. During the twentieth century, meteorologists greatly improved their skill in weather forecasting for relatively short time periods, such as out to a week. Their forecasts could be relied on by those making decisions about a wide variety of activities. Historically, however, public and private decisions about the weather over longer time periods—the climate—were based on a statistical analysis of weather records. Much has changed over the last two decades. Our ability to monitor and predict variations in climate has increased substantially. Growing knowledge of the causes and characteristics of seasonal to interannual variability and decadal-time-scale changes is being translated into useful long-lead-time forecasts and a greater capability to project future climate change. Those increases in capability coincide with a growing realization of the importance of climate variability and the potential for future change. The extension of climate information from a statistical analysis of historical observations to seasonal and interannual forecasts to century-scale projections has enabled a broader set of applications, which serve to enhance economic vitality, manage risk, protect life and property, promote environmental stewardship, and assist in the negotiations of international treaties. The array of applications and the potential for new applications are enormous.

The provision of climate services in the United States is evolving in response to the combination of a growing knowledge base, a growing appreciation of the importance of climate in human endeavors, and a greater demand for climate information. Two issues emerge. First, the value of these services to society depends on many factors, including the nature of uncertainties of the forecasts and projections, the strength of the connections between climate and specific human endeavors, the accessibility of the information, and the ability of users to respond to useful information. Second, the importance of climate is clearly stimulating user demand and broadening the scope of and demand for climate services. For this reason, the Federal Committee for Meteorological Services and Supporting Research asked the Board on Atmospheric Sciences and Climate (BASC) of the National Research Council to review the status of climate services in the United States and to recommend direction for the future provision of these services. That request matched the list of topics identified by BASC as priority subjects for study after its 1998 review of the field of atmospheric sciences. The federal agencies responsible for climate services developed a statement of task, and in response the board reviewed the provision of climate services in the United States.

The report that follows is divided into four main segments. Chapter 1 provides specific examples that show the growing breadth and demand for climate services in the United States. The evolution of climate services is examined in Chapter 2 by briefly reviewing its history and the current activities of U.S. agencies. The board developed a set of principles that should guide the provision of climate services; these are presented in Chapter 3. Finally, the board's recommendations focus on enhancing existing institutional capabilities. The recommendations, laid out in Chapter 4, reflect a set of "first steps" to a more effective climate service in the United States. The intent of the recommendations is to promote a climate service that is increasingly user-centric, that reflects the value of both historical and predictive knowledge, and that promotes active stewardship of climate information. The board believes that its recommendations can be implemented quickly and that they have the potential to pay large dividends at modest cost because they leverage current climate service efforts. At the same time, the guiding principles should continue to help promote a highly valued and useful climate service function in the United States even beyond these first steps.

Eric J. Barron, *Chair*
Board on Atmospheric Sciences and Climate

Contents

EXECUTIVE SUMMARY 1

1 INTRODUCTION 8
Climate Services Definition, 13
Examples Demonstrating the Breadth and Demand for Climate Services, 15
Summary, 22

2 EVOLUTION OF CLIMATE SERVICES IN THE UNITED STATES 23
Context, 23
Role of Government Agencies: A Bit of History, 24
Development of the Private Sector, 28
A Possible Future, 29

3 GUIDING PRINCIPLES FOR CLIMATE SERVICES 32
The Five Major Guiding Principles, 32

4 FIRST STEPS TOWARD AN EFFECTIVE CLIMATE SERVICE 39
Concluding Remarks, 56

REFERENCES 58

ACRONYMS AND ABBREVIATIONS 61

BOARD MEMBERS' BIOGRAPHIES 63

APPENDIXES
A Statement of Task, 71
B Workshop Participants, 72
C Agenda, 74
D Examples of areas of climate information requests, 77

EXECUTIVE SUMMARY

Climate[1] is an increasingly important element of public and private decision making in fields as varied as emergency management planning for increased threats of hurricanes or severe storms and energy-production requirements for the coming season. Advances in capabilities to monitor and predict variations in climate, coupled with growing concern over the potential for climate change and its impact, are yielding an increased awareness of the importance of climate information for enhancing economic vitality, maintaining environmental quality, and limiting threats to life and property (Changnon 2000).

Historically, climate services have revolved around the statistical analysis of existing weather records. Today, however, the science and understanding of global and regional climate have gone well beyond a statistical analysis of historical records. Research efforts of the last two decades have produced substantial improvements in understanding short-term climatic fluctuations such as El Niño and La Niña (NRC 1996). A number of groups around the world now regularly forecast aspects of the El Niño/Southern Oscillation phenomenon

---

[1] As weather forecasts for longer lead times have become possible and shorter term climate forecasts have been developed, there has been a blurring of the distinction between weather and climate. When used in this report, unless noted otherwise, weather forecasts are limited to two weeks in projection, and climate outlooks extend from two weeks and beyond.

and its associated impacts. Increasingly, climate model projections are also being used by decision makers to assess long-term global change issues of importance to the nation. After reviewing the various meanings used in the past, the National Research Council's Board on Atmospheric Sciences and Climate (BASC) took a broad view and defined *climate services* as the timely production and delivery of useful climate data, information, and knowledge to decision makers.

A climate service must focus on very different types of activities in order to address all the major categories of variability and change. Climate services and products include observations, forecasts, and projections and their uncertainties that address both seasonal to interannual variability and decadal-to century-scale change and variability, including human-induced global change. Each is associated with different types of users or decision makers and with different types of needs and products, as is evident by the current use of climate information.

The value of climate information to society depends on many factors, including the strength and nature of the linkages between climate, weather, and human endeavors; the nature of the uncertainties associated with climate forecasts; the accessibility of credible and useful climate information to decision makers; the ability of users[2] and providers to identify each other's needs and limitations; and the ability of users to respond to useful information. Increasing realization of the importance of climate is stimulating user demand for improved information, which in turn is substantially broadening the scope of climate services. Because of these factors, the subject of climate services was an agenda item at the fall 1999 BASC meeting, held jointly with the Federal Committee for Meteorological Services and Supporting Research (FCMSSR). Subsequent to the meeting, the federal agencies, through FCMSSR, asked BASC to review the status of climate services and to recommend direction for the future provision of climate services to the nation. In particular, BASC was asked to address the following items outlined in the statement of task:

- Define climate services.
- Describe potential audiences and providers of climate services.

---

[2] This term is used throughout the report to include various sectors of society that use weather and climate information. It includes various economic sectors (e.g., agriculture, transportation, industry, and insurance) as well as emergency managers, the media, and the general public.

- Describe the types of products that should be provided through a climate service.
- Outline the roles of the public, private, and academic sectors in a climate service.
- Describe fundamental principles that should be followed in the provision of climate services.

This report summarizes BASC's response to the task statement. BASC reviewed the subject at its spring 2000 meeting, at which it planned the details of a workshop that was held August 8–12, 2000, in Woods Hole, Massachusetts. Following the workshop, BASC dedicated a full day of its fall 2000 meeting to final deliberations and report preparation.

BASC assessed the current climate service activities in the United States and compared the state of climate services with the development of weather services to identify the following five guiding principles for climate services:

1. The activities and elements of a climate service should be user-centric.
2. If a climate service function is to improve and succeed, it should be supported by active research.
3. Advanced information (including predictions) on a variety of space and time scales, in the context of historical experience, is required to serve national needs.
4. The climate services knowledge base requires active stewardship.
5. Climate services require active and well-defined participation by government, business, and academe.

RECOMMENDATIONS

The goals of the following recommendations are to enhance the capabilities of existing institutions and agencies and to build a stronger climate service function within this context. Therefore, these recommendations constitute the "first steps" that can be taken immediately to enhance the effectiveness and efficiency of U.S. climate services rather than to reorganize existing activities. These first steps are designed to develop and provide climate services that are user-centric, that reflect the value of both historical and predictive knowledge, and that promote active stewardship of climate information. Taking these steps will pay large dividends at relatively modest cost[3] because several of the elements that are needed for climate services already exist. Furthermore, such recent advances in technologies as the Internet, data storage, and computing make possible economies that could not have been realized even a few years ago.

1. **PROMOTE MORE EFFECTIVE USE OF THE NATION'S WEATHER AND CLIMATE OBSERVATION SYSTEMS.**

Recommendation 1.1: Inventory existing observing systems and data holdings. Each agency should identify its climate-related observing systems and data holdings. For each observing system, the agency should identify (1) what purpose each set of observations and data serves for the provision of climate services, (2) how each observing system addresses user needs, (3) how each system is managed, and (4) what considerations govern decisions regarding the observing systems. The Office of the Federal Coordinator for Meteorology (OFCM) should be considered as an agent for this recommendation.

Recommendation 1.2: Promote efficiency by seeking out opportunities to combine the efforts of existing observation networks to serve multiple purposes in a more cost-effective manner. Strong interagency leadership is essential for creating a cost-efficient and cost-effective observing system. Again, this seems to be an appropriate activity for the OFCM.

Recommendation 1.3: Create user-centric functions within agencies. A truly useful climate service depends on having mechanisms that support and enable user-centric design and improvement. These include effective means for encouraging dialogue with users, the willingness to adapt existing data types and formats to meet specific user needs, and the availability of expertise in the operations of the various major user groups being supported within the climate service organization.

---

[3] No attempt to quantify the actual cost of implementing these recommendations was undertaken by BASC. The term *relatively modest* was used in contrast to some of the costs associated with recent NRC reports containing specific recommendations with respect to climate observing systems, modeling centers, etc. (NRC 1999b; 2001a).

Recommendation 1.4: Perform user-oriented experiments. A partnership of providers and users should be empowered to propose and execute experiments designed to promote and assess the use of climate information.

Recommendation 1.5: Create incentives to develop and promote observation systems that serve the nation. Currently across the nation there is a wide disparity in efforts to establish local-level weather and climate networks that augment stations run by federal agencies and are established to aid local and state decision-making.

2. IMPROVE THE CAPABILITY TO SERVE THE CLIMATE INFORMATION NEEDS OF THE NATION.

Recommendation 2.1: Ensure a strong and healthy transition of U.S. research accomplishments into predictive capabilities that serve the nation. The United States has a strong atmospheric and oceanic research community. However, there is a need to enhance the delivery of products useful to society that stem from this investment in research.

Recommendation 2.2: Expand the breadth and quality of climate products through the development of new instrumentation and technology. New instrumentation and technology should be viewed in terms of the expected expansion of the forecasting/prediction family into the areas of air quality, hydrology, and human health. To support new modeling and analysis capabilities and to support and improve the existing climate database, it is necessary to continue to improve upon existing sensors and their instrumentation and to develop new ones.

Recommendation 2.3: Address climate service product needs derived from long-term projections through an increase in the nation's modeling and analysis capabilities. There is a need to support a stronger role by the nation's modeling and analysis centers in the climate services related to long-term prediction. In particular, the centers should play a role in the development of the capabilities required to provide long-term simulations, analyses of limitations and uncertainties, and specialized products for impact studies.

Recommendation 2.4: Develop better climate service products based on ensemble climate simulations. There is a need for ensemble seasonal to interannual forecasts and climate simulations based on multiple emission scenarios that can be devoted to studies of climate impacts, vulnerabilities, and responses. This will require dedicated resources for developing ensemble climate scenarios, high-resolution models, and multiple emission scenarios for impact studies.

3. **INTERDISCIPLINARY STUDIES AND CAPABILITIES ARE NEEDED TO ADDRESS SOCIETAL NEEDS.**

Recommendation 3.1: Develop regional enterprises designed to expand the nature and scope of climate services. The nation should develop a program of regional entities (laboratories or centers) that emphasize region-specific observation, integrated understanding, and predictive capability to provide useful information that will drive the development of a regional focus on addressing societal needs.

Recommendation 3.2: Increase support for interdisciplinary climate studies, applications, and education. It is essential to provide support to foster both the capacity for making and the ability to beneficially use climate products that are based on data, information, and knowledge from many disciplines (e.g., combining physical, chemical, biological, and societal stressors[4] to yield products that explore climatological variability and societal impacts).

Recommendation 3.3: Foster climate policy education. Universities should initiate majors and minors in climate policy to enable informed planning and management of climate services. These programs should include education in the basics of climate science, identification of the needs of various user communities, and the creation and dissemination of climate information in forms that will be most useful to those users.

Recommendation 3.4: Enhance the understanding of climate through public education. Climate is increasingly important in decisions made by individuals, corporations, cities, states, and the nation as a whole. Critical to the successful application of climate information is an educated user.

---

[4] A term used in *Our Common Journey* (NRC 1999c) to describe anything that places a stress on the earth system, such as climate extremes, weather extremes (droughts and heat and cold waves), land use changes, and population excesses.

EXECUTIVE SUMMARY

BASC did not explicitly explore a formal climate services organizational structure within a specific federal, state, or local agency. Several such proposals internal to the government have been made in the past. One of note is the *NOAA Climate Services Plan* (Changnon et al. 1990), which offers several suggestions for consolidating the National Oceanic and Atmospheric Administration climate organizations, the Climate Prediction Center and the National Climatic Data Center, with the regional climate centers (RCCs) to form a unified climate service.

The existing network of state climatologists, RCCs, national agencies, and private sector organizations has provided services in the past and provides increasingly competent services today. BASC believes that the principles discussed in this report represent the best practices of the various activities and that if applied across all levels of service—local, state, regional, and national—would improve the overall climate services to the nation. The recommendations contained in this report offer concrete first steps toward a better integrated national system.

1

INTRODUCTION

In January 2020, the U.S. National Climate Service's annual outlook predicted much higher than normal probabilities for substantially reduced precipitation over the agricultural southeastern United States, as well as a continuation of the drought over the far West. The grain belt throughout the Midwest was projected to have near normal to above normal rainfall. The latest climate outlook from the European Climate Center agreed in general with the U.S. projection. The outlooks also predicted increased probabilities of much warmer than normal temperatures along the East Coast. Both the U.S. and European outlooks included a range of possible conditions and described the basis for the level of uncertainty. The outlooks included a historical context derived from the mining of the data archives. Analyses of the historical duration of similar past climatic events, their impacts, and successful mitigation strategies were incorporated as part of the risk evaluation process.

The annual outlooks serve as the basis for a wide array of planning activities and decision making. The utilities industries expanded their ability to support the increased power requirements of the elevated air conditioning demand that the summer was expected to bring and used a model of the distribution and timing of energy demand to purchase fuel during periods of anticipated reduced demand and hence lower price. The Department of Agriculture evaluated crop projections with respect to overall food demands by the United States and the nation's ability to support exports to other countries. With the expectation that drought throughout much of Africa would continue, the Department of State and the Department of Defense examined options for assistance that would be needed if social stresses increased throughout most of the continent. The Federal Emergency Management Agency (FEMA) issued warnings in regions prone to increased weather risk with the goals of improving mitigation efforts and increasing public awareness. U.S. industries guided the manufacture, marketing, and distribution of weather-sensitive products from clothing to home improvement goods and services. Travel agencies and the transportation industry planned for the expected demand for travel to specific winter and summer vacation destinations.

The hypothetical scenario above represents a potential for climate services two decades into the future if aspects of current capabilities and agency responsibilities are simply extrapolated forward in time. It illustrates the importance and impact of such a service, without even considering major breakthroughs in science, technology, or information management. The timely delivery of useful products through direct and accessible user interfaces can maximize the societal benefits and limit national risks. It represents directed efforts to translate knowledge of the climate system into a national service function.

Climate is an increasingly important element of the public and private decision-making process. Advances in monitoring and predicting variations in climate, coupled with growing concern over the potential for climate change and its impact, are yielding an increased awareness of the importance of climate information for enhancing economic vitality, maintaining environmental quality, and limiting threats to life and property (Changnon 2000). The importance of climate information has fostered the concept of a "climate service".

Climate has a local- to regional-scale component, and information and expertise are needed to address these space scales. For example, the climate of urban St. Louis is markedly different from that of rural Missouri, and the climate of the Midwest is strikingly different from that of the High Plains and the Northeast. Most activities impacted by climate operate at such local to regional scales and so need information for those areas. To serve the many agricultural interests in the Corn Belt, the Midwestern Regional Climate Center has developed an operational on-line climate-crop yield model. Users can call in any time during the corn (or soybean) growing season and get up-to-date information on the status of climate conditions for their area of concern—say, southeastern Indiana. They can then examine a variety of climatologically-based choices for weather conditions during the remainder of the growing season. And finally they can obtain crop-yield estimates based on what has already occurred and what might occur.

Historically, climate services have revolved around the statistical analysis of existing weather records. This information formed a useful basis for estimating future agricultural production and energy needs. Information on extreme events, such as "100-year floods" and peak wind speeds, continues to be critical in the design and construction of major facilities such as dams, highways, and buildings. In 1922, for example, the division of water allocations between states in the Colorado River Basin (the Colorado River Compact) was based on historical observations for a time of anomalously high flow, compared with later years. The allocations were based on a ten-year period to even out natural variability but could not incorporate longer time-scale variability. The Mexican Water Treaty of 1944 controls the allowable salinity and volume of water entering Mexico from the United States. The system of dams constructed along the Colorado River allows storage of four times the average annual flow as a hedge against low-flow years and as a means to maintain treaty requirements. Historical climate records are also used to set "safe" levels of water storage and prevent emergency water releases that would cause downstream flooding (e.g., Rhodes et al. 1984). The availability of the various sources of climate information and products summarized in this example has historically been supported through many federal, state, and local agencies and represented a climate service to the nation. In recognition of the growing importance of this service, the National Oceanic and Atmospheric Administration's (NOAA) National Climatic Data Center (NCDC) in Asheville, North Carolina, was founded in 1951 as the U.S. Weather Records Center.

Today, however, the science and understanding of global and regional climate have gone well beyond a statistical analysis of historical records. Research efforts of the last two decades have resulted in a substantial improvement in our understanding of short-term climatic fluctuations such as El Niño and La Niña (NRC 1996). A number of groups around the world now regularly forecast aspects of the El Niño/Southern Oscillation (ENSO) phenomenon and its associated impacts. In many regions, the impacts of ENSO are increasingly anticipated by decision makers as a result of forecasts that are taken seriously and serve as input for decision strategies and choices (Changnon 2000).

In April 1997, record flooding along the Red River of the North and in Grand Forks, North Dakota, resulted in $2 billion in damages. The magnitude of the damages was especially surprising given that the record flood was anticipated a season in advance. In February 1997, the U.S. National Weather Service issued a flood outlook for Grand Forks for the coming spring flood season warning residents to expect a flood of record, 49 feet, sometime in mid-April. While the 3-month outlook was within 10 percent of the actual flood crest—highly skillful by any measure of predictive accuracy—neither forecasters nor local residents fully understood the degree of uncertainty in the flood outlook (e.g., as to the timing and height of the water crest, the duration of the highest water, and the area covered by the flood). Consequently, preparations to fight the flood placed an improper precision on the 49-foot outlook, leading to surprise and then anger when the river's crest exceeded that level (Hooke and Pielke 2000).

The Grand Forks experience and many similar experiences send a two-pronged message. On the one hand, they provide optimism that forecasters can produce skillful predictions and other information on seasonal (and longer) time scales. On the other hand, they suggest that the effective use of the information is not always straightforward. As knowledge of climate and the use of climate information have developed, both the challenge and promise of providing operational "climate services" have become readily apparent. It is time to focus on the challenge of providing climate services to realize the promise that they hold.

The combination of historical observations, paleoclimate data, and efforts to develop a hierarchy of coupled models is increasing the understanding of the causes and character of climate fluctuations, although the nature of decade to century climate variability remains an important topic of research (NRC 1995, 1998c). From such research, a growing appreciation has emerged that "climate" can no longer be considered stationary—climate changes, sometimes dramatically, over periods of years to decades. Within the last two decades, concern about the potential of human-induced changes in the earth's climate has resulted in the development of new methods for observing the earth system and in better understanding of the possible courses of future climate. These developments and advances are in large part the result of core research in the atmospheric, oceanic, and biological sciences and are the research foci of the U.S. Global Change Research Program (USGCRP 2000). The increased understanding of the climate system has provided greater knowledge of the variations recorded by historical measurements. The tools developed for experimental long-lead-time forecasts (a month to a year) now permit experimental predictions of seasonal to interannual climate patterns and a limited but growing capability to project future climate changes. The societal value of such information depends on many factors, including the following:

- The strength and nature of the linkages between climate, weather, and human endeavors.
- The nature of the uncertainties associated with climate forecasts.
- The accessibility of credible and useful climate information by decision makers.
- The ability of users and providers to identify each other's needs and limitations.
- The ability of users to respond to useful information.

Increasing realization of the importance of climate is stimulating user demand for improved information and substantially broadening the scope of climate services. Because of these factors, the subject of climate services was an agenda item at the fall 1999 meeting of the Board on Atmospheric Sciences and Climate (BASC), held jointly with the Federal Committee for Meteorological Services and Supporting Research (FCMSSR). Subsequent to the meeting, the federal agencies, through FCMSSR, asked BASC to review the status of climate services and to recommend directions for the future provision of climate services to the nation. In particular, BASC was asked to address the following items outlined in the statement of task:

- Define climate services.
- Describe potential audiences and providers of climate services.
- Describe the types of products that should be provided through a climate service.
- Outline the roles of the public, private, and academic sectors in a climate service.
- Describe fundamental principles that should be followed in the provision of climate services.

This report summarizes BASC's response to the task statement. The board's efforts included a discussion of potential topics at its fall 1999 meeting; organizational discussions and preliminary briefings at its spring 2000 meeting;

an information-gathering activity during a workshop held August 8–12, 2000, in Woods Hole, Massachusetts; and additional discussion and review of materials at the fall 2000 meeting that followed the workshop.

CLIMATE SERVICES DEFINITION

Unlike climate services, the concept of a "weather service" is familiar to most citizens. A weather service focuses on the description, analysis, and forecasting of atmospheric motions and phenomena on very short times scales, extending up to a period of one week to ten days. The objective is to provide forecasts of continually changing weather and warnings of severe weather events. The benefits of this service are measured in lives saved, injuries avoided, or reduction in property damage. The National Research Council (1998a) has noted that this service is provided by a four-way partnership in which (1) the government acquires and analyzes observations and issues forecasts and warnings, (2) the government, newspapers, radio, and television participate in the dissemination of weather forecasts and severe-weather warnings, (3) private meteorological firms use government data and products to provide weather information for the media and special weather services for a variety of industries and activities, and (4) government, university, and private-sector scientists work to develop improved understanding of atmospheric behavior and turn it into new capabilities.

In contrast to weather, climate is concerned with the longer-term statistical properties of the atmosphere–ocean–ice–land system. Climate refers to the many statistical properties of such variables as temperature and precipitation over a specific region, the range of values of these variables, and the frequency of particular events as a function of location, season, or time of day. Climate variability and change are products of external factors, such as the sun; complex interactions involving the different components of the earth system; and human-induced (or anthropogenic) changes to the earth system. Following the World Climate Research Programme, BASC (NRC 1998a) adopted three general categories of climate variability and change: seasonal to interannual variability, decadal to century climate variability, and changes in climate induced by human activities, such as emission of greenhouse gases. A climate service must focus on very different types of activities to address all of these major categories of variability and change. Each of the activities is associated with different types of users or decision makers and with different types of needs and products, as is evident by the current use of climate information.

The earliest climate services consisted of descriptive information and statistical analyses of weather observations. The statistical analyses of the weather data evolved in response to the needs of engineers, designers, and insurers. Seasonal to interannual forecasts developed over the last few decades have generated additional demands for information and products. The value of the information is a function of the ability of the various sectors (e.g., agriculture, water, energy, transportation, and health care) to use climate information a season to a year in advance. The assessments of long-term potential changes in the global and regional climates as a result of anthropogenic factors add to the demands and variety of potential climate services. This information is being used by national and international policy makers and in a variety of long-range planning activities.

BASC reviewed the evolution of climate services in the nation (see Chapter 2) and examined the various definitions of *climate services* that have been used in the past. In reviewing the breadth of useful climate information—from historical analyses to century-scale predictions—BASC chose to be inclusive by adopting the following definition of *climate services*:

> The timely production and delivery of useful climate data, information, and knowledge to decision makers.[5]

---

[5] The term *decision maker* is intended to be generic, including anyone who uses climate information in the decision process—government or business planners, small business persons, farmers, the general public, etc.

EXAMPLES DEMONSTRATING THE BREADTH AND DEMAND[6] FOR CLIMATE SERVICES

The demand for and use of climate services has grown substantially over time. For example, NCDC now provides 120 major data sets and products on-line that describe a variety of climate elements. Since the initiation of the on-line data system in January 1998, orders have steadily increased to a rate of 1,800–2,000 requests per month. A profile of off-line orders reveals a customer base of about 77 percent business, 13 percent government, and 7 percent individuals (Karl 2000). Products of the NOAA Climate Prediction Center (CPC) include both extended weather outlooks and climate information. The most popular of the climate products include an ENSO Diagnostic Advisory, 30- and 90- Day Outlooks, long-lead-time (out to 13 months) outlooks, a U.S. Drought Assessment, and an Atlantic Hurricane Outlook. Users of NOAA's specialized data products include FEMA; the American Society of Heating, Refrigerating, and Air-Conditioning Engineers (ASHRAE); the Nuclear Regulatory Commission; and the National Association of Home Builders (NAHB). For example, NAHB uses temperature data when revising building codes. The air-freezing index can be used in conjunction with soil properties and surface ground cover to compute the ground freezing potential for particular areas and guide codes on building foundations. According to a U.S. Department of Housing and Urban Development study (HUD 1993), the revised building codes could result in annual savings of $300 million in building costs. ASHRAE is using NCDC temperature, humidity, wind speed, and solar insolation data with the goal of reducing user energy consumption by 50 percent. The U.S. Navy now uses monthly and seasonal climate outlooks to control the large heating systems on its installations, resulting in a savings of $65 million in utility costs (Cuff 2000). Previously, the dates for turning on the

---

[6] The term *demand* is not used in this report in the strictest economic sense. The government provides the basic weather and climate services in this country, and so the usual demand–price relationship does not exist. The many *customers* of weather and climate services express their needs and desires for information that will improve their operations. For example, the emergency manager of a coastal community *demands* greater accuracy in a hurricane forecast (both in real time and climate expectancy) to enable better performance in the planning for and execution of evacuation of the community as a storm approaches. Several examples of these expanding needs or *demands* are discussed in this chapter, but for a more detailed discussion of the many users of weather and climate services and their needs, see NRC 1998a, 1999d, and 2000a.

heating and cooling systems were determined using the average climate conditions for a particular region.

The practical use of climate information extends well beyond the long lead time (13 months) of the NOAA products described above. For example, the Navy has entrained long-term climate-model projections with Arctic sea ice models in its long-range planning for fleet appropriations and construction. Other enterprises with long lead times—including construction of offshore drilling rigs, planting of timber, and development of transportation options—have also included climate change projections in their decision making.

A wide variety of cases demonstrate positive outcomes or reduced risk when climate information is incorporated into the decision-making process. However, because a comprehensive survey of such cases is not possible, this report cannot assess the magnitude of the value and importance of climate services. Instead, four examples have been selected from those investigated by BASC to illustrate of the breadth and character of the growing demand for climate services. The examples were selected to show the expansion of climate services from statistical analyses to more sophisticated products, including prediction as well as the range of partnerships involved in providing and accessing climate services. The first two examples, (1) coastal exposure and community protection from severe storms and (2) weather derivatives, describe services built directly on analyses of historical observations but for which there is a growing demand for enhanced predictive capability. The third example, based on ENSO forecasts, directly examines the importance of seasonal to interannual forecasts. The fourth example, based on the recent U.S. assessment of future climate impacts, considers the value of longer-term climate projections. In addition to showing both the breadth of and growing demand for climate services, the examples guide several of the recommendations in this report. Many more examples could be cited as well, involving, for example, water-resource management, water quality–climate connections, droughts and agriculture, and human health.

1. Climate information for the control of risks in coastal communities. Historical averages of weather data have been used in many applications throughout the nation. For example, the historical record shows that more hurricanes hit southern Florida than any other part of the country, prompting that region to write the nation's most stringent building codes. However, Hurricane Andrew, which crossed southern Florida in 1992 as a powerful category 4 storm, dramatically changed the risk and vulnerability perceptions of residents, governments, builders, and insurance companies. FEMA's Federal Insurance Administration has identified outdated codes and a wide variety of inadequacies in construction, code enforcement, inspection, and training. Insurance industry experts estimated that up to 40 percent of the insured damages could have been avoided if building codes that were developed using more accurate climate information had been enforced (Pielke and Pielke 1997). The fact that damages were much greater than expected (an estimated $15.5 billion in insured damages and total damages about twice that) has also resulted in revised estimates of coastal damage potential, with numerous eastern seaboard cities' insured-loss projections exceeding $50 billion for a category 4 or 5 hurricane (Insurance Research Council 1995). Substantial savings can be realized if building codes that take improved climate data into account are adopted by builders and enforced. However, an ability to anticipate risk and vulnerability also has potential value.

Both societal exposure (coastal population growth and wealth) and the climatological understanding of the frequency and intensity of weather hazards are of major importance to the insurance industry (e.g., Munich Re 1994, Berz 1993, Dlugolecki 1992). The industry cannot provide coverage blindly and therefore must base the price and scope of underwriting on an assessment of potential liability. The potential liability of the industry has typically been based on a statistical analysis of past events to determine the probability of future events and on a geographic assessment of a variety of various liability categories. Those factors are combined to determine loss ratios. Both likely loss and probable maximal loss are important. Over the last several decades there has been significant growth in insured losses. Typically, changes in premiums and in scope of underwriting do not occur until after a substantial payment of claims. However, the increased insured losses have prompted the insurance industry to examine all aspects of its liability, including those dealing with changing climate patterns—either natural variability or longer-term climate change. The key issues identified by the industry are how weather changes with seasonal to decadal variations, whether climate change will be associated with new extremes, and whether changes in weather patterns due to seasonal to decadal variability or climate change will result in changes in the geographic location of exposures or the intensity or frequency of events. Hence, the role of climate services has been extended from a statistical analysis of the historical record to one in which understanding of climate variability and change in the context of population and demographic shifts becomes of substantial importance to governments and industry. The insurance industry, particularly the reinsurance portion of the industry, seeks better climate information and outlooks so it can better manage its exposure. In addition, the insurance industry has established, as a priority, programs to mitigate the loss of life and property. Improved climate information assists the industry in planning for those important activities.

2. **Weather Derivatives.** Weather derivatives are financial instruments designed to allow businesses sensitive to the vagaries of weather to protect themselves against loss from changes in costs and sales linked to variations in climate. They gained in popularity during the winter of 1998–99 when long-range forecasts called for warmer than normal weather during that period (due to the onset of El Niño). Weather derivatives can theoretically be designed for almost any weather variable (e.g., rain, snow, and wind), although most of the activity so far has been based on temperature. They are quite different from insurance in that insurance is based on the willingness of individuals or corporations to pay a premium to transfer risk. The premium is based on weather statistics and a potential damage function, and the payout from the insurance reflects actual damages incurred. However, for many enterprises weather risk is defined by its impact on performance rather than on loss of property (e.g., impact on production rates, product availability, pricing, and product demand). The number of industries whose performance is impacted by weather and climate is large (WeatherRisk 2000). For example, moisture availability is a key factor in agriculture; under adverse growing conditions, the largest corn-producing counties in the nation can experience 30 percent reductions in yield. Agrochemical industries are also impacted; for example, weevils cost cotton growers $300 million annually, and weevil severity is directly correlated with mild winter conditions. Viticulture has a strong climatological signal, with a lack of sunshine and cool temperatures from pre-bloom to maturation producing poor grape years and excess precipitation causing grape rot. The 1998 California grape harvest was 30 percent lower than the norm because of poor weather conditions. Beverage consumption and ice-cream sales are strongly correlated with warm temperatures. Clothing retailers are recognizing correlations between summer and winter sales volume and the magnitude of the departure from normal weather conditions. Amusement parks' profit margins are defined by peak attendance days associated with good weather. The construction industry often experiences penalties of 10–15 percent of job costs for delays related to, for example, temperatures needed for concrete to set,

wind conditions, or rain-freeze events. Weather derivatives are designed to address such examples of climatic impacts on performance.

The key to the growth of a weather derivatives market is that weather risk has differential impacts in terms of costs and benefits—a two-way flow of risk promotes the concept of trading risk. High snow accumulation aids a ski resort by increasing profits, but it substantially increases the cost of snow removal in cities. Hot summer conditions can increase power demand by a factor of 20 at greatly increased costs. Heat stresses livestock and agriculture, but it aids the beverage, ice cream, and summer clothing industries. Large fluctuations in performance associated with weather and the differential nature of impacts are fostering a weather derivatives industry. Weather derivatives add stability because during optimal conditions profits are high, so paying for the derivatives is not a burden. During poor conditions, the payout from derivatives helps to maintain cash flow and profitability, so risk is managed.

The first transaction in the weather derivatives market took place in 1997. In 2000, between 2,500 and 3,000 weather-related contracts were issued in the United States, with a total value of $2.5 billion (PricewaterhouseCoopers 2001). The Weather Risk Management Association commissioned a recently completed survey by PricewaterhouseCoopers to determine the exact state of the weather derivatives industry (see <http://www.wrma.org>). EnronOnline (<http://www.enrononline.com>), an Internet weather derivatives leader, currently has a contract portfolio of $5 billion. The largest economic sector involved is energy because of the strong and obvious links between temperature extremes and energy demand, but both the number and scope of companies have been increasing since 1997 (including such new fields as leisure and home improvement). Some changes that are occurring in climate services are helping the growth of the weather derivatives industry. These include the availability of NCDC databases; the formation of private climate service companies that provide clean, standardized data directly for weather risk assessment at a large number of locations; the development of stochastic modeling tools as an aid in risk analysis; and the development of operational seasonal forecasts at the NOAA National Centers for Environmental Prediction (NCEP), particularly involving ENSO (WeatherRisk 2000).

3. The applications of regional ENSO forecasts. The 1925–26, 1972–73, and 1982–83 El Niños, which resulted in dramatic changes in rainfall and flooding, are examples of short-term climate events that had substantial societal impacts. The magnitude of ENSO and its impacts, a growing observational and modeling capability, and a desire to better understand the role of the tropics in governing climate variability resulted in the development of the 10-year Tropical Ocean Global Atmosphere program. Important outcomes of that effort have been the development of a tropical observing system in the Pacific Ocean and some demonstrated skill in producing El Niño forecasts. NCEP's CPC is now experimenting with the use of coupled ocean–atmosphere forecasts, together with several statistical techniques, to arrive at an official consensus El Niño forecast with a lead time of two seasons. The scientific community is far from demonstrating skill in predicting all the features of ENSO (e.g., onset, duration, amplitude, and decay). However, once the El Niño is under way, the tendency for it to persist into the boreal winter results in substantial capability to anticipate and plan for many of its effects (Landsea and Knaff 2000).

Such investigations are just beginning in the United States, but such El Niño events as those of 1997–98 provide the opportunity to demonstrate the potential value of an increased ability to anticipate climate phenomena. The CPC correctly forecasted six months in advance the increased precipitation and storms across California, the southern plains, and the Southeast and a warmer winter in the northern tier states. An analysis of the losses and benefits associated with the 1997–98 event is instructive (Changnon 1999). In response to the 1997 climate predictions and on the basis of knowledge gleaned from the 1982–83 El Niño, FEMA initiated several mitigation activities, particularly in California and Florida. For example, California spent $7.5 million on flooding-landslide preparedness efforts and alerting the public, and the roofing and home repair industry reported $125 million in completed mitigation activities (Changnon 1999). The U.S. Geological Survey and the Bureau of Reclamation used the forecasts to guide water management in western systems. A number of northern utility companies used the forecasts to guide oil and natural gas purchases. For example, three utilities in Iowa and Michigan saved $39, $48, and $147 million through fuel purchase planning (Changnon 1999). Although it is difficult to estimate, 161 fatalities and $2.2 billion in losses were attributed to the 1997–98 El Niño in the United States (Changnon 1999). Of note, the losses in California were estimated (Changnon 1999) to be half those of the 1982–83 El Niño, suggesting that the extensive mitigation efforts were extremely beneficial. The impacts of ENSO variability and the potential savings resulting from the effective use of forecasts of ENSO events cover many areas, including agricultural production, coastal fisheries, emergency preparedness, systems for early warning of droughts, strategies for hydro-electric power generation, insurance concerns, and mitigation of forest fires (NRC 1996).

4. **International and national assessments of the impacts of future climate change.** The Intergovernmental Panel on Climate Change (IPCC) was established by the World Meteorological Organization and the United Nations Environment Programme to assess the science of climate change. The IPCC is charged with generating an assessment of the state of knowledge of climate change every five years. A critical product of this process is the *Summary for Policymakers*, which details the elements of the assessment deemed most significant for national and international decision makers. The IPCC has become a standard by which governments assess the implications of climate change, and IPCC reports underpin international negotiations on climate change and carbon emissions. In the *Summary for Policymakers*, the IPCC identifies the importance of the maintenance and improvement of the global observing system, the monitoring of key climate elements, the incorporation of prehistorical data into the examination of the climate record, the development of comprehensive climate models, and focused process studies as the foundation for advice to policy makers (IPCC 2001). The United States, as directed by the Global Change Research Act of 1990, has also undertaken a National Assessment of Potential Consequences of Climate Variability and Change, a landmark effort to examine the potential consequences for the regions of the nation, including coastal regions, and for forestry, water, agriculture, and human health. The historical record and the results from global climate models act as the foundation for stakeholders to assess the implications of climate change and then to consider adaptation strategies (USGCRP 2001).

Unlike earlier reports, the National Assessment does not analyze the potential for climate change. Instead, climate information is the basis for examining consequences of and vulnerabilities to climate change. A number of the consequences have important economic and quality-of-life implications. Climate modeling and analysis are the foundation for developing climate scenarios that describe alternative futures for analysis of potential consequences of climate change, potential adaptation options, and ultimately the vulnerability of communities, institutions, sectors, and regions. In addition, there is a call for greater emphasis on sustained quality-controlled observations. The assessment process reflects a growing need by decision makers and stakeholders for future climate information. In this important context, the climate observing system and the national capability to provide future climate projections represent a climate service to the nation (NRC 1998b, 2001b; USGCRP 2000).

SUMMARY

Climate information is increasingly important to the decision-making needs of a wide variety of users. The range of applications is enormous and indicates the impact of climate services. Applications include such diverse entities as water set-asides for instream flows to protect ecosystems, state and national water compacts, building codes, insurance premiums, irrigation and power production decisions, beverage consumption rates, retail clothing volumes, and construction schedules. The applications are also growing with the increased understanding of how climate influences human endeavors. That a weather derivatives industry, designed to manage risk associated with climate variability, is growing rapidly and expanding to include diverse elements of commerce and industry is indicative of the importance of climate services to the nation. The extension of climate capabilities from a relatively straightforward, statistical analysis of historical observations to seasonal and interannual forecasts and to century-scale projections has enabled a broader set of applications, including enhancing productivity in weather-sensitive industries, managing weather risk, protecting life and property, and negotiating international treaties.

Public, private, and academic sectors have all played important roles in the extension. Much of the innovation for individual industry and economic sector use has been developed in the private sector. The academic community continues to improve understanding of climate variability and predictability through its research. The public sector has improved the accessibility of its data and information, including the output from the extensive forecast models now being run. Further, the scientific and user communities are increasingly articulating the improvements required in climate services to enable improvements in decision making.

2

EVOLUTION OF CLIMATE SERVICES IN THE UNITED STATES

CONTEXT

The provision of climate services in the United States is continually evolving in response to a growing understanding of climate, from the scales of natural variability to long-term climate change, combined with a growing appreciation of the intersection of climate and human endeavors. The context for this evolution is best viewed by starting with a brief historical review and then looking at the present National Weather Service; the state climatologist program; nascent ocean observing, modeling, and predicting endeavors; and other federal, state, and local activities, with a vision of a more comprehensive "environmental service". This context arises because "climate service" can be considered as a (thoughtfully) structured portion of a larger applied earth sciences program with a greater degree of organization, purpose, and continuity that is likely to result in substantially greater national benefit. The weather service began with a system that was based largely on observations and, as capabilities increased and demands for service were more clearly articulated, expanded to a predictive service. Climate services have an analogous evolution, with an observation-based historical emphasis followed by a growing emphasis on abilities to forecast seasonal to decadal changes. While it is analogous to a weather service in that sense, it is important also to acknowledge the major differences.

ROLE OF GOVERNMENT AGENCIES: A BIT OF HISTORY

Changnon (1995) describes one of the earliest applications of climate data as the use of weather observations in colonial times to plan agriculture and construction. The U.S. Weather Bureau, founded in 1870, established the Weather Records Center, later to be called the National Climatic Data Center (NCDC) in 1951. The name was a clear indication of the concept of climate information in the 1950s. The next major evolution in the provision of climate services was the Weather Bureau's creation of the state climatologist program in 1954 (Hecht 1984). The program was an attempt to better link the needs of state and local users and the capabilities of the Weather Bureau. The program was successful in some states and not particularly successful in others. Because of budget pressures, uneven results, and increasing concerns over excessive federal involvement in what was perceived to be a state role, the program was terminated in 1973, and states were left to determine what kind of state climatologist program, if any, would be supported. The present situation is also mixed. The current president of the American Association of State Climatologists divided the state programs (including that of Puerto Rico) into four categories. Seven states have what he defined as good programs, 19 have satisfactory programs that are good but limited by funding, 21 are marginal, and 4 have no program at all (Angel 2000). One noteworthy example of a successful state program is the Oklahoma Climatological Survey (OCS) (see Box 2-1). Such successes provide important lessons for extracting "best practices" for efforts to move toward more effective climate services in the United States.

Box 2-1
The OCS and the Oklahoma Mesonet:
A Climate Services Success Story

The OCS was established in 1982 by legislative mandate to "acquire, archive, process and disseminate, in the most cost-effective way possible, all climate and weather information that is or could be of value to policy and decision makers in the state." It has developed into an operation with a budget of $3 million per year that employs 30 full-time staff and 20 students. In addition to the usual historical data records and studies contained in typical state programs, the OCS developed the Oklahoma Mesonet (funded in 1991), a detailed observing network of 114 sites across the state with 3,100 sensors used to identify smaller scale weather and climate regimes and to provide detailed information to users. The OCS provides a variety of data products and access to models linked to weather and climate information (e.g., fire damage, dispersion for pesticide application, evapo-transpiration, and pest and plant disease). Easy user access to the information is the goal. The OCS received over 3.5 million requests in 1999. Access is provided on the Web as a self-service operation for television, news services, universities, and area businesses. Call-in requests are handled for a variety of stakeholders—from farmers to insurance agents to attorneys. The OCS is founded on strong partnerships with research programs at several federal agencies (the Department of Energy, the Department of Agriculture, the National Science Foundation, the National Aeronautics and Space Administration, the National Oceanic and Atmospheric Administration, and the U.S. Army Corps of Engineers) and a host of state agencies (such as those related to water resources, environmental quality, forestry, civil emergencies, and agriculture). Outreach is a significant part of the charter and a K–12 program, a public safety outreach program, and a rural electric outreach program are formally established. The OCS charges for non-Oklahoma and for-profit data requests to supplement state funding and to ensure that Oklahoma taxpayers do not underwrite the requests of non-taxpayers. The director, Ken Crawford, attributes the success of the OCS to the following: (1) users have been involved from day one, (2) products are developed in direct partnership with users, (3) strong partnerships exist with mission agencies and with research elements, (4) information is accessible, and (5) education of users and potential users is an important element of the program (the OCS holds workshops on such topics as correct interpretation of weather and climate information). By honoring the mandate of ensuring decision makers access to information of value, the OCS has become a success story. In 1999, it was selected by Harvard University and the Ford Foundation as one of the twelve "most innovative American government programs." (Crawford 2000).

In response to the increasing need for climate information, Congress passed the National Climate Program Act of 1978. The purpose of the act was to promote the understanding of climate and the provision of climate services. The act directed the development of a network of regional climate centers (RCCs) to meet regional climate service needs (Hecht 1984). The centers were established over the ten-year period 1981–90. The executive branch, however, never embraced the concept of the RCCs because they were all created by congressional earmarks and add-ons; the Department of Commerce proposed their elimination each year, and Congress restored them in each appropriations cycle. Six RCCs currently exist and report to NCDC, which also maintains close ties to the state climatologists. The Western Regional Climate Center processed 35,000–40,000 requests from 1982 to 1995, and the demand is increasing. The center's characterization of the requests demonstrates interest in a broad array of topics, such as agriculture, engineering, legal issues, and economic development (see Appendix D).

The RCCs' focus is to apply climate information to regional issues, recognizing the "regionality" of the issues facing decision makers. The centers emphasize the breadth of climate service requirements, including evaluation of data, archiving, indexing, retrieval, quality assessment, synthesis, interpretation, and dissemination, and maintain a central focus on the user (Redmond 2000). A major climate product of the RCCs, and one not available at NCDC or most state climate offices, is on-line access to daily updated climate conditions, including the temperature and precipitation conditions of yesterday (or the last few days) and interpretations of how those conditions rate climatologically against values from the past 50–100 years. That type of information is often sought by those attempting to rate what has just occurred against past behavior and past decisions. For example, government decision makers often seek such information to get a rating on recent extremes, such as a recent two-day rainstorm that produced flooding in southern Wisconsin. Regional centers identify a key problem in providing climate services—that is, climate services are made up of a mixture of elements or components without effective national integration. The National Oceanic and Atmospheric Administration (NOAA) has been unsuccessful in its attempts to link the activities of NCDC, the RCCs, and the state climatologists.

The Climate Analysis Center, established in response to the National Climate Program Act, is a part of the National Weather Service's (NWS) National Centers for Environmental Prediction. When first established, it included a diagnostics component, a data and information component, and a prediction component. The climate diagnostics component received considerable attention after the 1982–83 El Niño. With the rapid scientific advances in the understanding of the El Niño/Southern Oscillation that took place in the 1980s and early 1990s, the center's name was changed to the Climate Prediction Center (CPC). Recognizing that a major service now provided by the CPC is the seasonal to interannual forecasts issued for the United States, the NWS recently created a Climate Services Division within its headquarters to provide oversight and direction of the CPC.

The climate activities within NOAA now consist of the CPC within the NWS, NCDC within the National Environmental Satellite, Data, and Information Service, and several research laboratories[7] within NOAA's Office of Research.

The National Aeronautics and Space Administration's (NASA) mission includes efforts to support the development of new and innovative observations, research, modeling, and data and information management that have a direct impact on the potential to provide climate services. NASA and NOAA provide satellite and in situ observations that are the backbone of the climate observing system. Mechanisms for establishing sustained long-term climate satellite and in situ observations, which are viewed by the Board on Atmospheric Sciences and Climate (BASC) as essential to supporting a wide variety of climate services, are yet to be realized (see NRC 2000b, 2000c). NASA has also developed an applications program to create public-private partnerships that result in the use of NASA science and technology to improve environmental, community growth, resource management, and disaster management decisions. The NOAA and NASA efforts to combine forecasts and observations through data assimilation and reanalysis are providing an invaluable best estimate of a host of weather and climate variables that are difficult to measure. This reanalysis using state-of-the-art models is rapidly becoming the de facto climate record.

A number of the agencies participating in the U.S. Global Change Research Program[8] (USGCRP) contribute to the development of climate services through data collection and/or efforts to improve climate prediction capa-

---

[7] Among the NOAA laboratories that produce widely used products are the Aeronomy Laboratory with its trace-gas data, the Climate Modeling and Diagnostics Laboratory with its observatories at Pt. Barrow, Mauna Loa, American Samoa, and South Pole, and the Climate Diagnostics Center with numerous climate diagnostic products.

[8] The USGCRP does not include all the federal agency activities that support climate, particularly climate services. For example, the environmental satellites of NOAA are not included in the USGCRP even though they contribute significantly to the climate record.

bilities. For example, the U.S. Department of Energy maintains a series of research observation sites that contribute to the understanding of climate processes. It has also supported climate modeling development through its carbon dioxide research programs. The National Science Foundation has provided support for improved climate model parameterizations and increased understanding of natural variability such as ENSO. The U.S. Environmental Protection Agency has a growing effort to link climate change with the assessment of impacts on water, including water quality, and human health and a continuing interest in maintaining observation capabilities to assess environmental standards for air and water. The U.S. Geological Survey (USGS) and some states maintain hydrologic measurement stations across the nation. Many USGS gauging stations are supported by joint federal-state funding. The U.S. Department of Agriculture (USDA) maintains a Joint Agricultural Weather Facility that relates crop production to weather and climate events; this facility is staffed jointly by the USDA and the NWS (USGCRP 2000).

The nation has also provided substantial support to increase the ability to model climate variability and climate change through a commitment to fund national centers, including the National Center for Atmospheric Research, the NOAA Geophysical Fluid Dynamics Laboratory, and the NASA Goddard Institute for Space Studies. Such efforts are critical in providing climate services for national assessments of climate impacts, for participating in such international assessment activities as the Intergovernmental Panel on Climate Change (IPCC), and in considering U.S. participation in international treaties. However, those institutions were designed as research centers, not as service centers. In addition, the United States has provided support for such innovative institutions as the International Research Institute (IRI), whose mission is to show that forecasts on seasonal to interannual time scales can be used successfully in real and meaningful applications.

Many of the research aspects of the various agency programs are coordinated through the USGCRP and are reported annually to the nation through the publication of *Our Changing Planet* (see, e.g., USGCRP 2000). However, there is no comparable mechanism for promoting or coordinating the climate services aspects of the various agencies.

DEVELOPMENT OF THE PRIVATE SECTOR

After World War II, thousands of meteorologists who had been trained for the war effort returned to the United States. Many went back to their previous employment; some went into the U.S. Weather Bureau and academic institutions; but several formed private weather forecasting services and began applying the skills they honed in wartime. The latter and their companies responded to the needs of weather-sensitive businesses and industries for weather and climate information that was more tailored to their specific needs than what could be provided by the Weather Bureau.

In 1957, the American Meteorological Society, recognizing the need for a professional certification process, initiated the Certified Consulting Meteorologist (CCM) program. The growing need for specialized weather- and climate-related services has resulted in substantial growth in the environmental consulting industry and in the demand for trained meteorologists. There are currently over 550 CCMs.

At present there are over 250 private weather firms listed on the NWS Web site (<http://www.nws.noaa.gov/im/more.htm>), and most offer climate services as part of their product suite.

A POSSIBLE FUTURE

For the last 60 years the United States has developed and supported a national weather service that today uses vast amounts of meteorological data, assimilates these data into models of the atmosphere, and produces and disseminates about 24,000 weather forecasts each day (Hooke and Pielke 2000). At the beginning of the twenty-first century, two changes are occurring. First, growth in knowledge is enabling recognition of the interactions of a multitude of environmental stresses, including land-use change, climate variability and change, waste products, population growth, and health and well-being. Second, technological advances in earth observing sensors and systems, in computers and information storage, and in communication networks are enabling a much more comprehensive approach to environmental conditions. Combined, those two factors provide the nation with the potential to transform a weather service with an inherently short-term perspective into a decision-centric environment service that spans the spatial scales of local to global concerns and the temporal scales of minutes to a century, including both weather and climate (NRC 1999c).

Aided by that technological evolution, a broader, organized, and sustained environment service will probably emerge as the needs of decision makers become more complex. In addition, the environmental approach should reflect the fact that most regions are influenced by multiple stresses, such as weather and climate, land-use change, and input and character of pollutants. *Our Common Journey: A Transition Toward Sustainability* (NRC 1999c) recommends moving toward a research framework that looks at all elements that influence a specific locale or region (multiple stresses). An environmental service will likely develop as this more holistic place-based approach to earth sciences is undertaken. Such an approach was also recommended in *Grand Challenges in the Environmental Sciences* (NRC 2001c) and *The Science of Regional and Global Change: Putting Knowledge to Work* (NRC 2001b). In this view, climate service is an important ingredient in a service that will ultimately support integrated management. It is a logical and needed next step.

The climate services under discussion in this report would use observations of the physical, chemical, biological, and geological state of the solid earth and its surface cover, the ocean, and the atmosphere extending from the earth's surface to outer space. The observations would include the meteorological, hydrologic, oceanographic, atmospheric, and space observations currently made. Central to the scope of climate services is the need for them to embrace wide ranges of time and space scales. This is an expansion of what the NWS uses as a data base and predictive framework, but it is not as comprehensive as an environment service, in which the data and predictive models would include, for example, species populations and distributions, and the data systems would be modified to provide the accuracy and stability necessary for climate.

In many ways, the evolution of technology and growth in knowledge have enabled the establishment of a climate service and will continue to foster the evolution of a climate service to an environment service. Data storage and exchange have been made possible. Computational resources suitable for such an endeavor are within reach. Many of the global observing capabilities are now online in the ocean, as well as on land and in space. Increasingly, global and regional atmospheric and oceanic models are moving toward understanding the variability and predictability of the fluid envelope of the earth system, including its biogeochemical, hydrologic, and ecosystem elements. Those capabilities are reflected in the scope of the climate service in this discussion. The present emphasis on assessments at global to regional scales sets the stage for this new climate service. The NWS, the IRI, the IPCC, the U.S. Assessment, the RCCs, and the state climatologists can be viewed as prototypical, and such activities are likely to evolve in response to the demand for climate information. However, BASC concluded that a more cohesive management and integration will result in a comprehensive climate service that is capable of integrating and synthesizing data from diverse sources and is willing to accept the charge to do so, rather than a diverse set of services for different elements of the earth system. The climate service should be inherently integrative and synthetic, capable of bringing together diverse data sets and advancing the ability to understand and predict climate variability and change. It should produce products and achieve the benefits of access to global data and global participation.

3

GUIDING PRINCIPLES FOR CLIMATE SERVICES

Climate services, by definition, are mission-oriented and driven by societal needs to enhance economic vitality, maintain and improve environmental quality, limit and decrease threats to life and property, and strengthen fundamental understanding of the earth. At an August 2000 workshop, the Board on Atmospheric Sciences and Climate (BASC) reviewed the various climate service activities described in the previous chapters. Representatives of the various activities (the state climatologists' programs, the regional climate centers, the National Climatic Data Center, the Climate Prediction Center, private sector organizations, etc.) were asked to discuss not only their activities, but also those characteristics that make their activities successful or present difficulties. The discussion led to the identification of best practices that are suitable for an overall climate service for the nation. Those best practices were then distilled into five guiding principles and are set forth below. The principles were developed through an assessment of current climate service activities and substantial experience within the atmospheric sciences in the development of weather services.

THE FIVE MAJOR GUIDING PRINCIPLES

1. **The activities and elements of a climate service should be user-centric.** The interface between the knowledge base and the user is critical if

the objective is the timely production and delivery of climate information relevant to the user's decision needs. "Climate forecasts are useful only to the extent that they provide information that people can use to improve their outcomes beyond what they would otherwise have been." That statement, from *Making Climate Forecasts Matter* (NRC 1999a), is part of a series of findings that describe the importance of the interface between users and data providers. The report and climate services examples cited earlier are the basis of a set of requirements for a user-centric climate service:

- A comprehensive service should strive to meet the needs of a user community at least as diverse and complex as the climate system itself, ranging from the international community to individual users and involving both the public and private sectors. Central to the scope of a climate service is the need to embrace wide ranges of time and space scales because decision making occurs on all scales from local to global and from weeks to centuries.
- Users will become increasingly diverse, knowledgeable, and specialized. Consequently, their needs will evolve. Greater education of users in the meaning and significance of climate information is likely to promote greater use and more robust application of the information.
- The key to an effective climate services program is a vigorous, cost-effective, and comprehensive intersection of knowledge and its use. Therefore, the following elements are essential for a successful program:
  - Mutual information exchange and feedback.
  - Communication and accessibility of information.
  - Continuing evaluation and assessment, by users and providers, of the use and effectiveness of the services.

2. If a climate service function is to improve and succeed, it should be supported by active research. The ability to serve national climate needs is a direct product of the U.S. investment in the pursuit of new and useful knowledge. A continuous and concerted effort to develop and incorporate new knowledge is a requirement of any sustainable service.

- Research should focus on improved understanding of the dynamics of the diffusion of knowledge and information, including how it is transferred, communicated, and used and the implications of its use.
- Climate services should be an objective of mission-oriented research.
- Active mechanisms should be employed to enable the transition from research discovery to useful products (NRC 2000a).

3. **Advanced information (including predictions) on a variety of space and time scales, in the context of historical experience, is required to serve national needs.** The case studies in Chapter 1 describing the current demand for climate services illustrate the scope of the knowledge base that should be included to serve national needs. The knowledge base should include the following:

• Continuous, accurate, and reliable historical climate observations at diverse locales. Many applications use time series of climate variables to estimate trends, departures from average conditions, and extremes (low-probability events).

• Access to climate observations that include the perspective of the paleoclimatic record where available and appropriate to guide understanding of natural variability.

• Forecasts and outlooks, from a month to a year in advance, including an analysis of probabilities, limitations, and uncertainties and mission-oriented specialized products, which are of major importance to a large segment of the user community. Understanding the skill of such products is critical for their effective and beneficial use.

• Access to the growing knowledge base on the causes and character of natural variability on time scales of seasons to decades.

• Long-term climate simulations, starting from the beginning of the last century when the robust historical record begins and extending to the next 100 years. Such simulations provide an important service to the nation when coupled with an analysis of limitations and uncertainties.

• Information on spatial scales ranging from local to regional to global.

4. **The climate services knowledge base requires active stewardship.** The quality, consistency, accessibility, and documentation (including limitations and uncertainties) of climate information are a ubiquitous concern of current users and help to define requirements for stewardship of the nation's climate information:

• Open and free exchange of data is essential to incorporate the interdependence and teleconnectivity of elements of the global earth system and to consider the full strategic interests of the United States as part of the global economy.

• Reliable long-term observations and data archives are of critical importance (see Box 3-1).

- Emphasis on converting observations into useful records is essential. Multi-purpose observations, including in situ and space-based systems, and an array of technologies and variables have proved to be the most useful for the characterization of climate.
- The most successful climate observing systems have included synergism among observations, theory, and modeling.
- Development and maintenance of a robust and easily accessible delivery system are necessary for an effective climate service.

> **Box 3-1**
> **Climate Monitoring Principles**
>
> The U.S. climate community has long articulated the basic principles for monitoring climatic variables. Using a one-year assessment of atmospheric observations and data management, the National Research Council's Climate Research Committee provided recommendations for how the National Weather Service modernization might address the needs of the climate community (NRC 1992). Its nine "principles of observing and managing data for climate and climate change research" were subsequently expanded and articulated as "Ten Basic Climate Monitoring Principles" (Karl et al. 1995). The ten principles were subsequently endorsed by the National Research Council report *Adequacy of Climate Observing Systems* (1999b):
>
> - *Management of network change.* Assess how and to what extent a proposed change in the observation network will influence the information from the system.
> - *Parallel testing.* Operate the old system simultaneously with the replacement system over a sufficient time period to connect climatic data taken before and after the change.
> - *Metadata.* Fully document each observing system and its operating procedure. Documentation should be carried with the data.
> - *Data quality and continuity.* Assess data quality and homogeneity as part of routine operating procedures. This assessment should focus on the requirements for measuring climate variability and change.
> - *Integrated environmental assessment.* Anticipate the use of data in the development of environmental assessments, particularly those per-

taining to climate variability and change, as a part of the observing system's strategic plan.

- *Historical significance.* Maintain operation of observing systems that have provided homogeneous data sets over a period of many decades to a century or more. Develop a list of key protected sites, based on their prioritized contributions to the long-term record.
- *Complementary data.* Give the highest priority in the design and implementation of new sites or instrumentation within an observing system to data-poor regions, poorly observed variables, regions sensitive to change, and variables with inadequate temporal resolution.
- *Climate requirements.* Give network designers, operators, and instrument engineers climate monitoring requirements at the outset of network design. Instruments must have adequate accuracy to resolve climate variations and changes of primary interest. Use modeling and theoretical studies to identify spatial and temporal resolution requirements.
- *Continuity of purpose.* Maintain a stable, long-term commitment to these observations, and develop a clear transition plan from serving research needs to operational purposes.
- *Data and metadata access.* Develop data management systems that facilitate access, use, and interpretation of data and data products by users. Freedom of access, low-cost mechanisms that facilitate use (directories, catalogs, browse capabilities, availability of metadata on station histories, algorithm accessibility and documentation, etc.), and quality control should be integral parts of data management.

5. Climate services require active and well-defined participation by government, business, and academe. BASC reviewed the development of the partnership between the public and private sectors in the provision of weather services. The partnership was codified by the National Weather Service (NWS) in 1991 with a formal policy statement that replaced the long-standing, unwritten policy (NWS 1991). Many scholars have addressed the public and private roles over the years, most recently Stiglitz et al. (2000). Stiglitz, who served as chief economist of the World Bank and as chairman of the President's Council of Economic Advisors, reviewed the balance of activities between the public and private sectors in various areas, including an explicit look at the NWS. The criteria presented in this report are consistent

with the analysis therein. This long history helps to define the roles of the participants in the delivery of climate services:
- The role of all government agencies should be motivated by the concept of "public goods and services." Those are goods and services that are *non-rival* and *non-exclusive*. The following premises should guide these activities:
  - Tax dollars should not subsidize activities of individuals or individual commercial operations.
  - The government has responsibility for managing issues that are clearly in the wider public interest, including protection of life and property, increasing understanding of climate and weather, and improving services through basic and applied research.
  - The government has international responsibilities for implementing and coordinating programs on behalf of the public.
  - The government should be responsible for maintaining the nation's official climate records.
- The role of the private sector is motivated by and acts under market forces and owes principal responsibilities to its own and its clients' interests. This defines substantially different premises:
  - The private sector generates data and products to which it may retain a proprietary interest, even if the data are subsequently transferred to a public agency unless otherwise agreed upon.
  - The private sector may maintain the confidentiality of its dealings with its clients and hence of the content and character of the information it provides to its clients (subject to any applicable legal actions or restrictions).
  - The private sector has substantial freedom in determining the conditions (restricted or unrestricted) for providing its value-added data and services to the public, academe, or government agencies.
  - The private sector engages in basic and applied climate research to meet the needs of its users, concentrating on user-centric products and services.
  - Members of the private sector constitute a principal resource for the innovation, development, and manufacture of advanced technological devices and equipment for making climatological observations, for computing, and for telecommunications.
- Academic research organizations play a critical role in climate education and research. These activities are focused primarily on the following:

- Traditional university roles of research, education, and outreach, much of which may be funded by federal research and development agencies and state governments.
- Research, data compilation and analysis, and product development, sometimes administered as part of a hybrid academic/government culture (e.g., the National Center for Atmospheric Research, the National Oceanic and Atmospheric Administration/University Cooperative Research Institutes, and offices of state climatologists).
- Research, data compilation and analysis, and product development in partnership with industry (including technology commercialization) that also fulfill the research, education, and outreach mission of academe.

BASC provides these principles for consideration in future climate service organizations, whether they are entirely new organizations or a refocusing or realigning of existing organizations. BASC believes that these best practices, if applied universally, could improve the provision of climate services to the nation.

4

FIRST STEPS TOWARD AN EFFECTIVE CLIMATE SERVICE

The scope and importance of climate services in the United States is growing. That growth is a logical result of increased capabilities to monitor, understand, and predict climate variations; increased concern about the potential for future climate change; and increased awareness of the value of climate information. Climate information is and will be used in many diverse ways to support decision making, and a climate service should reflect this diversity. Its scope, therefore, should be defined by the temporal and spatial scales that make climate information useful.

The value of weather information is in its timeliness, so its value decreases quickly with age. In contrast, a substantial number of the climate-related user requests described in this report are based on the analysis of time series of variables to estimate trends, departures from average conditions, and extreme conditions (low-probability events). Climate services have a product orientation that extends from weekly to centennial time scales. Therefore, the value of climate observations tends to increase with accuracy, consistency, and continuity over time. However, the sampling interval must be short (e.g., to capture diurnal variability) to create an accurate statistical basis for analysis of both short-term extremes and long-term trends. Accurate knowledge of weather-scale variability is essential for producing climate products. Historical records add context to the understanding of variability. The scope of the supporting modeling endeavors should be equally comprehensive. Models become increasingly valuable if they add value to the observational base, correctly anticipate departures from the norm a season or a year in advance, or help to define either risk or opportunity tied to longer-term trends. Therefore, the scope of the modeling efforts should include the use of model-data hybrids to create long-term climatologies of variables that are not directly observed, a combination of statistical and dynamical models to assess conditions a season to a year in advance, and coupled earth-system models designed to incorporate changes in the factors (e.g., greenhouse gases, aerosols, solar variations, and land-cover change) that force long-term changes to the climate system (NRC 1998b, 1999b).

The range of spatial scales is equally diverse. Climate variability can have small spatial scales, and many climate products will be place-based (specific to one site). Hence, high resolution becomes a critical need in data used in creating site-specific products and in developing gridded products to guide decisions. Where inputs and products cannot be portrayed on common high-resolution grids, the ability to use models to downscale information is required to provide the requisite high-resolution products. Although many problems are site specific, the generation of climate products will rely on data not only across disciplines and time but also across space to points distant from the place of interest. For example, a regional seasonal forecast of precipitation and temperature in the United States will rely on ocean observations and surface marine observations, as well as on soil moisture, to initialize the global coupled ocean–atmosphere–land model used to produce the forecast. At the same time, downscaling of the forecast requires local climatologies, statistics, topography, land-use data, and other local or regional information.

As the decisions vary over space and time, climate services should at once be responsive on the local level and integrative of all the wide-ranging and diverse influences on that place. Fundamental to the development of climate information that serves the needs of the nation is a commitment to a global observing system (NRC 1999b); recognition of the importance of place-based, local and regional observations; a strong service-oriented modeling capability (NRC 1999c; 2001a); and a commitment to a user-centric focus (NRC 1999a).

The observing system required for the climate services is global. At the same time, the place-based studies of climate impacts will often require a higher resolution of observations. Therefore, attention to the various national and regional networks is also required. In response to changing capabilities and needs, the infrastructure of climate services cannot be seen as static. An important aspect will be the design and optimization of the observing system. That should be undertaken through the interplay of models and observations and requires validation of models (comparison with observations) and user feedback. Although the synthesis of observations by assimilation into models cannot replace the need for a comprehensive suite of observations, the design of the observing system should be undertaken in the context of ever-improving models. Thus, observing system simulation experiments that can guide technological advances for observations will be an important component of climate services (NRC 2000a).

It is also essential that the United States be engaged in, through its climate service, the free and open exchange of data and products and that it benefit from the participation of non-U.S. scientists (NRC 1995). Climate variability increasingly points to the interdependence and teleconnectivity of elements of the earth system on a global basis. The economic well-being of other countries is often in the strategic interest of the United States; hence, national interests to be served by a climate service should be broadly defined (NRC 2000a).

The temporal (time scale) scope of observations, forecasts, and projections mandates on the one hand that climate services be closely linked to the present activities of weather services and on the other hand that it develop a capacity to support IPCC-type assessments routinely. One important challenge of initiating a climate service involves understanding the nature of the climate assessment process, defining and institutionalizing the active role that a service should play in the process, and building in the flexible, ever-changing interfaces that should be maintained among the multiple players who will perform along the pathway from data to decision. The process of adaptive learning (or the integration of learning and action) in this enterprise should be anticipated in the institutional setting used and should be expected to continue indefinitely. Because the service will be defined in large part as a decision-support system, it is important that it be developed by focusing on a broad suite of practical problems that will illuminate the breadth and character of stakeholders. Among the products of the climate service should be those that quantify the uncertainties in climate forecasts and analyses. The service should make clear to users the sensitivity of the products to the data and methods used in preparing them. Such products, ongoing self-evaluation within the climate service, and user feedback will be important in guiding the evolution of the climate service (NRC 1999a).

The following recommendations are constructed around a strategy that is based on enhancing the capabilities of existing institutions and agencies and building a stronger climate services function within this context. Therefore, the objective is to define the first steps that can be taken immediately to enhance the effectiveness and efficiency of U.S. climate services. These first steps are designed to promote climate services that are user-centric, that reflect the value of both statistical and predictive climate knowledge, and that promote active stewardship of climate information. The Board on Atmospheric Sciences and Climate (BASC) did not assign explicit priorities to the recommendations, but lower-numbered recommendations generally are more important than higher-numbered recommendations in the same section. Taking these first steps will pay large dividends at relatively modest cost because several of the elements that are needed for climate services already exist. In addition, such recent advances in technologies as the Internet, data storage, and computing make possible economies that could not have been realized even a few years ago.

1. **PROMOTE MORE EFFECTIVE USE OF THE NATION'S WEATHER AND CLIMATE OBSERVATION SYSTEMS.**

Recommendation 1.1: Inventory existing observing systems and data holdings. The climate observing system is the backbone of any climate service. A fully integrated observing system that supports climate services, either for climate prediction or the provision of regionally tailored climate products, does not currently exist. No agency currently has responsibility for carrying out or coordinating a comprehensive program of climate observations (NRC 1998d). The National Research Council report (2001b) *The Science of Regional and Global Change: Putting Knowledge to Work* summarizes this key issue: "The observing 'system' available today is a composite of observations that do not provide the information needed nor the continuity of the data to support decisions on many critical observations." A number of federal agencies, including the National Oceanic and Atmospheric Administration (NOAA), the Department of Agriculture (USDA), the National Aeronautics and Space Administration (NASA), the Department of Energy (DOE), the Federal Aviation Administration (FAA), and the Department of the Interior (DOI), operate observing systems that could be components of a fully integrated climate observing system. A first step in such an integration is an inventory of existing observing systems, their data holdings, and their management rules. BASC could find no evidence of such a basic inventory of observations and their management. Therefore, the board recommends that each agency identify its climate-related observing systems and data holdings. The inventory should include information identifying the following: (1) what purpose each set of observations and data serves for the provision of climate services, (2) how each observing system addresses user needs, (3) how each system is managed, and (4) what considerations govern decisions regarding the observing systems.

The National Research Council report (1998a) *The Atmospheric Sciences Entering the Twenty-First Century* suggests that "the Federal Coordinator for Meteorological Services and Supporting Research should lead a thorough examination of the issues that arise as the national system for providing atmospheric information becomes more distributed." Following that statement and in consideration of the role of the federal coordinator, BASC suggests that the Office of the Federal Coordinator for Meteorology[9] (OFCM) be considered as an agent for this recommendation. To the extent practicable, the OFCM should include a survey of private or publicly funded but privately operated research programs or operational climate facilities. Additionally, each should be evaluated for adherence to the ten principles of climate observations (see Box 3-1), and estimates should be made of the resources necessary to bring them into full compliance. Initiatives for acquiring the resources should be a high priority in the budget process.

Recommendation 1.2: Promote efficiency by seeking out opportunities to combine the efforts of existing observation networks to serve multiple purposes in a more cost-effective manner. "Full interagency leadership is needed to create a cost-effective and balanced observing system" (NRC 1998a). The current multi-agency approach creates problems with both balance and effectiveness. For example, the current observational approach creates gaps because there is no long-term framework or funding for building integrated, sustained, end-to-end capability (NRC 2001b). That problem is evident in the number of cases in which the observational approach relies on capturing opportunistic measurements from research programs. Furthermore, observations that are currently made by many agencies are often driven solely by an agency's focused mandate, without consideration of low- or no-cost steps that could be taken to make the data more useful to a broader array of users

---

[9] The Office of the Federal Coordinator for Meteorological Services and Supporting Research (also called the Office of the Federal Coordinator for Meteorology (OFCM)) was established by the Department of Commerce in 1964 in response to PL 87-843. Its mission is to coordinate operational meteorological requirements, services, and supporting research among the federal agencies. A full description can be found at <http://www.ofcm.gov>.

(NRC 2001b). The inventory of existing observing systems recommended above is the first step in creating a more efficient system. The second step should be a deliberate effort to promote efficiency and balance. BASC recommends (1) examining the systems for redundancy and determining the merit of the redundancy, (2) examining potential synergisms between systems whereby co-location, co-management, or joint planning for future improvements would improve the knowledge base and/or save substantial costs, and (3) examining systems for gaps and weaknesses (including how data are managed) that could be addressed through more efficient operations or management. Following Recommendation 1.1, that seems to be an appropriate activity for the OFCM.

This recommendation reflects numerous examples of potential synergism between local, state, and regional networks, many of which have been set up to serve specific needs and are managed as such. For example,

- The 700-station USDA SNOTEL (Snow Telemetry) network is designed to furnish hydrologic and temperature information from mountainous locations in support of water supply forecasting for the western states. It is the only extant large-scale high altitude network in the nation (or the world) and could supply valuable information on climate variability and climate change in mountainous regions (the source of most of the West's streamflow) and greatly assist with regional reanalysis.
- The 15-year-old Remote Automatic Weather Station (RAWS) network is managed by DOI's Bureau of Land Management and USDA's Forest Service (and other agencies). The network records hourly meteorological data from 830–930 stations (depending on the season) throughout the western states. This data set is downloaded via satellite to the National Interagency Fire Center in Boise, Idaho, and immediately transferred for archival at the Western Regional Climate Center in Reno, Nevada. Originally oriented toward fire management, the network has increasingly important applications in natural resources management throughout the public lands of the West.
- Over the past decade, the FAA has installed over 1,000 automated observing systems at airports across the country. These systems, which are designed to provide aviation weather information, could contribute to improved climatologies of wind and temperature. Suitable upgrades to achieve the ten principles listed in Box 3-1 would increase the utility of these networks to serve the climate community without requiring the establishment of a new and expensive supporting infrastructure. Similar enhancements to other networks could lead to similar results. Workshops should be convened to

examine the issues involved in making such networks more efficient and to recommend priorities for such upgrades. Participants would include managers and providers of the networks and their data sets and potential customers for such information, from both the public and private sectors. A state-by-state analysis of all observational networks should be performed to analyze the suitability of existing networks in providing useful climate data and to consider possible upgrades. Data base links to a centralized climate service should be encouraged.

This process could be used to integrate the existing observational capability and to determine the adequacy of existing networks, the potential for their augmentation, and the gaps that remain in the climate observations. An option for addressing the gaps is proposed in Recommendation 1.5.

Recommendation 1.3: Create user-centric functions within agencies. An evaluation of the performance of the federal government in the 1997–98 El Niño identified the lack of clearly delineated agency or interagency functions designed to evaluate user demand and needs continuously, to create feedback that improves climate services through the interaction of users and producers, and to create new capabilities and functionalities that serve the needs of decision makers (Changnon 2000). That is but one example of the importance of the interface between the knowledge base and the user if the objectives are the timely production and delivery of climate information relevant to the decision needs of the user. The NRC (1999a) report *Making Climate Forecasts Matter* provides guidance on the nature of the interface: "Participatory approaches to delivering climate information might include structured dialogues between climate scientists and forecast users to identify the climate parameters of particular importance to users and the organizations that users might rely on for climate forecast information. . . . They would tend to make forecast information more decision relevant, to improve mutual understanding between scientists and forecast users, and to encourage appropriate interpretation and use of forecast information." In addition, the report argues for systematic efforts to bring scientific outputs and user needs together to increase the utility of forecasts (NRC 1999a).

The above reports, combined with the review of current and potential climate services by BASC, yielded the first guiding principle for climate services (see Chapter 3)—that is, the activities and elements of a climate service should be user-centric. Three elements, given in Chapter 3, are essential for a vigorous, cost-effective, and comprehensive intersection of knowledge and its use:

(1) mutual information exchange and feedback, (2) communication and accessibility of information; and (3) a continuing evaluation and assessment, by users and providers, of the use and effectiveness of the services. Those three essential elements are unlikely to occur if there are not user-centric functions within agencies. There must be formal mechanisms that support and enable user-centric design and improvement. BASC recognizes that this is not a simple task. There can be a strong cultural and linguistic gap between the climate researcher and the user. Effective involvement of the user community is essential.

Recommendation 1.4: Perform user-oriented experiments. A partnership of providers and users should be empowered to propose and execute experiments designed to promote and assess the use of climate information. The effectiveness of information depends strongly on the systems that distribute it, the channels of distribution, recipients' modes of understanding and judgment about the information sources, and the ways in which the information is presented (NRC 1999a). The research base for examining the effectiveness of information distribution and the value of climate information and forecasts through research on actual data use is thin (NRC 1999a).

The development of a comprehensive research program to examine the benefits and dissemination of climate information is outside the scope of this report. However, BASC found considerable uncertainty associated with the types and value of products that should be provided by different agencies as part of a climate service. A first step to addressing this concern is the design of experiments that would permit a rapid assessment of users' needs, wants, and preferences at low cost. For example, to test and demonstrate the utility and user acceptability of climate information data furnished in a real-time Web-based environment, the federal government could fund the user access charges for an already existing climate data network(s), allowing the user community free access to the data stream. That would provide the managing agencies with an almost instant statistical compilation of the user activities that include such elements as the number of users and the number and context of user accesses. Typical data sets to be considered might be the Oklahoma Climatological Survey Mesoscale Network, SNOTEL, and SCAN (Surface Condition Analyzer).

Recommendation 1.5: Create incentives to develop and promote observation systems that serve the nation. Currently across the nation there is a wide disparity in efforts to establish local-level weather and climate networks that augment stations run by federal agencies and that are established to aid local and state decision making. A substantial new program to coordinate and manage the nation's observing systems is probably unlikely in the near future, given the multiplicity of agencies and missions involved in collecting and disseminating information. However, initial steps that promote greater coordination; active stewardship involving quality, consistency, accessibility, and documentation of climate information; and open and free exchange of information will enhance our ability to serve the nation. After the results of implementing Recommendations 1.1 and 1.2 are assessed, additional observing capability might still be required. With this objective in mind, BASC examined best practices associated with current climate service functions. Oklahoma's mesoscale network is an example of an extensive network that helps the state to provide its citizens with a wide array of climatological services, such as public safety in the face of severe weather, assessment of drought, recommendations for weather- and climate-related agricultural management (e.g., spraying), and fire danger assessment. Other states have less extensive networks, such as Nebraska's Automated Weather Data Network (see <http://hpccsun.unl.edu/awdn/home.html> for a more comprehensive discussion of local networks), or no established network at all beyond the federal sites of the FAA or the National Weather Service (e.g., Pennsylvania).

A federal matching program for states and regional centers should be initiated to develop observation systems that obey the ten principles (in Box 3-1), promote free access, and create strong partnerships with users. Part of a set of initial steps that could be taken to establish a more comprehensive national climate network is the establishment of state or regional networks. Related local climate services should also be encouraged. One possible mechanism would be for the federal government to establish a matching program for states and/or regional climate centers to develop a climate observing network. The advantages of such a program are that it would be relatively inexpensive; it would enhance the value of the observations by following basic guidelines for accuracy, consistency, metadata, etc.; and it would allow the data network to be tailored to local climate problems.

2. **IMPROVE THE CAPABILITY TO SERVE THE CLIMATE INFORMATION NEEDS OF THE NATION.**

The science and understanding of global and regional climate have gone well beyond a statistical analysis of historical records under the assumption of climate stationarity. Forecasts of seasonal to interannual climate variations and long-term climate projections have become part of the breadth of climate service products. In many cases, the types and nature of the forecast or prediction products that should be provided through a climate service require additional efforts that promote a transition from research to operational efforts and new investments in modeling and modeling infrastructure. The recommendations in this section are derived from several recent National Research Council reports that outline needs for improvements in the climate observational and forecasting systems and for effective planning of the transition from research to operations. The recommendations given here use the findings of those reports as a means of emphasizing critical contributions and needs for improved climate services.

Recommendation 2.1: Ensure a strong and healthy transition of U.S. research accomplishments into predictive capabilities that serve the nation. The United States has a strong atmospheric and oceanic research community. This investment in the pursuit of new and useful knowledge has the potential to enhance our ability to serve national climate needs. In many ways, the research effort represents a significant body of new and useful products for a variety of decision makers. However, there is a need to enhance the delivery of products useful to society that stem from this investment in research. Many of the most important aspects of this enhancement are described in the report *From Research to Operations in Weather Satellites and Numerical Weather Prediction: Crossing the Valley of Death* (NRC 2000a). Additional discussions of climate-related transition issues are discussed in *Issues in the Integration of Research and Operational Satellite Systems for Climate Research, Parts I and II* (NRC 2000b, 2000c). The key elements include implementing development, testing, and integration capabilities to incorporate observations and advances in understanding into predictive models; providing adequate staffing for operational missions; providing a continuing process for assessing technology and updating it as needed to accomplish the mission; creating stronger collaborative efforts in the development of community prediction models; and institutionalizing the transition process from research to operations. Many of those recommendations were addressed to the Environmental Modeling Center at NOAA and operational sensor/satellite development between NASA and NOAA, but the basic themes can be easily transferred to other observation and predictive modeling efforts.

Recommendation 2.2: **Expand the breadth and quality of climate products through the development of new instrumentation and technology.** BASC's twenty-first century report (NRC 1998a) included two imperatives. The first was to develop a specific plan for optimizing global observations by taking into account the requirements for weather, climate, and air quality and for the information needed to improve predictive models. The second sought a commitment for developing new capabilities for observing critical variables, including water in all its phases, wind, aerosols, and chemical constituents. The twenty-first century report specifically notes the importance of a series of emerging issues (climate and health, management of water resources, and emissions of pollutants into the atmosphere) to the atmospheric sciences. A focus on those emerging topics increases the value of atmospheric information to society. Chapter 3 states that "A comprehensive service should strive to meet the needs of a user community at least as diverse and complex as the climate system itself." To support new modeling and analysis capabilities and to support and improve the existing climate database, it is necessary to continue to improve existing sensors and their instrumentation and to develop new ones. The key is to examine the new instrumentation and technology in terms of the expected expansion of the products and services (NRC 1998a; 1999d) and to develop capabilities that address their production in service to the nation. Air quality and hydrologic products are logical first investments. The types of specialized sensors that should be developed include some for measuring (1) air pollutants in order to couple weather, climate, and air quality and (2) soil moisture, radiation, elements of land use, and boundary-layer profiles of temperature, humidity, and wind on regional and subregional scales to better predict surface hydrology and moisture conditions.

The new instruments should obey the ten principles listed in Box 3-1 for climate observations that were adopted by BASC as part of the guiding principles for climate services. New technologies should have at least the same capabilities as the old, they should not be a burden on the research enterprise, and clear transition modes to operational use should be established (NRC 2000a).

Recommendation 2.3: **Address climate service product needs derived from long-term projections through an increase in the nation's modeling and analysis capabilities.** Climate modeling and analysis are the foundation for developing climate scenarios for the future. In turn, the scenarios are inputs for national and international decisions on potential adaptation and mitigation efforts. The U.S. National Assessment of Climate Change Impacts on the United States calls for a stronger capability for providing such climate information. The nation's climate modeling expertise is widely recognized as the best in the world, and this expertise is dedicated to developing state-of-the-science model capability. However, these research enterprises currently do not provide the ensembles of long-term simulations, extending from the start of the robust historical record to at least the next 100 years, that are required to serve national and international assessments. The Assessment states that "the demand for these climate services exceeds the capabilities of the research functions of the nation's climate modeling centers." In fact, the U.S. Assessment used Canadian and United Kingdom model products to examine the impacts of climate on the United States. Furthermore, the current U.S. National Assessment of Climate Change Impacts demonstrates a demand for a host of specialized climate products that tie future climate projections more directly to specific decisions or vulnerabilities. The assessment process requires greater access to and greater understanding of the limitations inherent in future projections to weigh the advantages and risks associated with alternative courses of action. An important requirement is the development of scenarios for the evolution of the factors that force climate (e.g., aerosols and greenhouse gases). Coupled system models, operating at substantially finer resolution, are required to link climate with the scales of human decisions. The nation's modeling and analysis centers need to develop the capabilities to provide long-term simulations, analysis of model limitations and uncertainties, and specialized products for impact studies (NRC 1998b, 2001a).

Recommendation 2.4: Develop better climate service products based on ensemble climate simulations. There is a need for ensemble seasonal to interannual forecasts and climate simulations that can be devoted to studies of climate impacts, vulnerabilities, and responses. The nation is scientifically capable of providing these products; however, it will require dedicated resources for developing ensemble climate scenarios, high-resolution models, and multiple emission scenarios for impact studies. Such an investment would enhance the capacity of the climate modeling community to generate and analyze model runs that are dedicated for use by impact analysts. Similarly, future assessments need to investigate a range of plausible emissions and atmospheric concentrations of carbon dioxide and other greenhouse gases. Enhancing the capability to generate dedicated scenarios of emissions and

FIRST STEPS TOWARD AN EFFECTIVE CLIMATE SERVICE 51

climate would dramatically improve the range of outcomes that future assessments of vulnerability could analyze (NRC 1998b, 2001a).

3. INTERDISCIPLINARY STUDIES AND CAPABILITIES ARE NEEDED TO ADDRESS SOCIETAL NEEDS.

Recommendation 3.1: Develop regional enterprises designed to expand the nature and scope of climate services. BASC's twenty-first century report (NRC 1998a) recognized the importance of the expansion of the forecast family into the areas of water, air quality, and human health to serve national needs. *Our Common Journey* (NRC 1999c) argues for a framework that integrates global and local perspectives to shape a place-based understanding of the interactions between science and society with the objectives of mitigating threats to society and meeting basic human needs such as providing food and nutrition and ensuring air and water quality. It calls for more interactive linkages between those who create knowledge of the earth system, together with technology development, and those who use the knowledge to support decision making. *The Science of Regional and Global Change: Putting Knowledge to Work* (NRC 2001b) articulates a set of action items to ensure an "intimate connection" between research, operational activities, and the support of decision making and for regional and sectoral multiple-stress research to build decision-support capability. The overall vision of *Grand Challenges in Environmental Sciences* (NRC 2001c) suggests that the key to future environmental research will be to develop the capability to examine regions comprehensively to address a broad range of environmental problems and issues.

Those reports argue for transitions in the ability of science to serve society. At the core of each report is the notion that robust observation, data management systems, and forecasting and predictive capability can be expanded to serve decision makers better. Climate observations and climate forecasting and projection provide a foundation for the expansion of the observing and forecasting family, and hence the expansion of the family of services and products. Chapter 2, which describes the evolution of climate services in the United States, articulates a possible future of climate services that evolves from a weather service with an inherently short-term perspective to decision-centric climate services and eventually to environmental services that include climate information in the broader context of multiple stresses and spatial scales that range from local to global concerns. The recommendations suggest that the development of climate services in the United States should include strategies that recognize that the future will bring a demand for services that focus on a multitude of problems rather than on specific disciplines.

A place-based imperative for environmental research stems from the importance of human activities on local and regional scales; the importance of multiple stresses, including climate, on specific environments; and the nature of the spatial and temporal linkages between physical, biological, chemical, and human systems. The strongest interaction between human activity, climate, and other environmental stresses is at the regional scale. Consequently, climate services, with a specific focus on regional problems and issues, can have their strongest intersection with decision makers. To address that need, the nation might begin to develop and fund a program of entities (laboratories or centers, either publicly, privately, or academically based) that emphasizes region-specific observation, integrated understanding, and predictive capability to produce useful information and focuses on addressing societal needs. Such laboratories or centers should be a part of the competitive, peer-reviewed research enterprise with the objectives of developing an integrated framework that will engender new avenues of research and application, catalyzing the development of useful products in service to society, bringing a demanding level of discipline to the full range of climate, and enabling investigation of a broad array of environmental issues through an enhanced capability at regional scales (NRC 1999c). These regional laboratories or centers are in many ways a regional subset of the objectives and guiding principles associated with climate services. As such, they should incorporate the elements of integrated observations, information systems, framework for supporting research, predictive products, and strong user interface described throughout this report:

Integrated observations. Current observations are often driven by different mission needs and tend to focus on the measurement of discrete variables at a specific set of locations. The observational systems are designed to serve the needs of weather forecasting, pollution monitoring, hydrologic forecasting, or other objectives. Recommendation 1.1 addresses the complex issues and challenges of creating integrated observing systems. However, a benefit of a regional focus is that the integration of complex sets of observations is manageable at this level, and the specificity of the problems and issues acts as an incentive for integration. That view is supported by current regional climate centers (RCCs) that demonstrate substantial value for a variety of users through the integration of a variety of data products. At a regional level, there is a potential to (1) link observing systems into a web of integrated sensors, building upon existing weather and hydrologic stations and remote sensing capability, (2) formulate the agreements across a set of more limited agencies and federal, state, and local governments needed to create a structure for the observing system, (3) provide a compelling framework that encourages or demands the integration of new observations into a broader strategy, and (4) create strong linkages between research and operational observations that result in mutual benefit. The result is likely to create new efficiencies through the development of more comprehensive measurement systems that are more useful in enhancing the capabilities of RCCs.

Regional information systems. Society has amassed an enormous amount of data about the earth. Fortunately, technological innovations are allowing the capture, processing, and display of this information in a manner that is multi-resolution and four-dimensional. The major challenges involve data management; the storage, indexing, referencing, and retrieval of data; and the ability to combine, dissect, and query information. The ability to navigate this information, seeking data that satisfy the direct needs of a variety of users, is likely to spark a new "age of information" that will promote economic benefit and engender new research directions and capabilities to integrate physical, biological, chemical, and human systems. A regional focus becomes a logical test bed, enabling the participation of universities; federal, state, and local governments; and industry in the development of a regional information system that is tractable and whose immediate benefit for a state or region can be evident.

Framework for process studies. Process studies are a critical element of scientific advancement because they are designed, through focused observations and modeling, to probe uncertainties in knowledge about how the earth system functions. In many cases, a mismatch between model predictions and observations can drive targeted investigations to limit the level of error. Frequently, efforts to couple different aspects of the earth system (e.g., the atmosphere and land-surface vegetation characteristics) prompt targeted exploration because the level of understanding is still rudimentary. The objective is to use field study to address deficiencies in understanding. The benefit of these intensive studies is maximized when they can be coupled with a highly developed, integrated set of sensors, with readily accessible spatial and temporal data within a regional information system and a predictive model framework that readily enables the entrainment and testing of new information from process studies.

Predictive capability. The value of reliable advanced information is widely recognized. Prediction is an important path for the translation of knowledge into economic benefit and societal well-being. Over the last several decades there have been substantial increases in the ability to forecast weather and project climate and climate variability into the future. Enormous potential exists if the present mesoscale models can be coupled with improved regional mesoscale observing systems. Such a capability enables a strategy and implementation capability for building tractable coupled models, initiating experimental forecasts of new variables, assessing the outcomes associated with multiple stresses, and taking advantage of the discipline of the forecasting process to create a powerful regional prediction capability. Built on the numerical framework of weather and climate models, this capability could be extended to air quality, water quantity and quality, ecosystem health, human health, agriculture, and a host of other areas. Even in areas where predictive capabilities are not yet matured, or not forthcoming, such knowledge would allow decision makers to structure their decision processes in ways that are more independent of predictive information.

User-centric functions. By creating an observation, process study, and predictive capability that addresses multiple stresses in specific regions, the opportunity can be created to develop research capabilities that are tuned to the needs of users. At this scale, it would be possible to incorporate stakeholders and their decision-making needs at the outset and create a vigorous, cost-effective, and comprehensive intersection with knowledge and its use. Education, communication, outreach, and a continuous assessment process should be key requirements for a successful regional research enterprise.

The regional vision described above is designed to address a broad range of current and future environmental issues by creating a capability based on integrated observing systems, readily accessible data, and an increasingly comprehensive predictive capability. With demonstrated success over a few large-scale regions of the United States, this strategy will likely lead to a national capability that far exceeds current capabilities and permit the creation of a broader class of environmental services.

Recommendation 3.2: Increase support for interdisciplinary climate studies, applications, and education. It is essential to provide federal support to foster both the capacity for making and the ability to beneficially use climate products that are based on data, information, and knowledge from many disciplines (e.g., combining physical, chemical, biological, and societal stressors to yield products that show climatological variability and societal impacts). That support should be developed as a direct element in support of mission agency objectives and as a part of the objectives of independent, exploratory research agencies. The existing climate-related observing, modeling, and service and product-providing capacity in the United States is a foundation on which to build, as is the existing federal, state, and local support funding infrastructure. However, those structures also present challenges. *Our Common Journey* (NRC 1999c) describes the need to integrate disciplinary knowledge in place-based, problem-driven research efforts and notes that this need runs counter to deeply held organizational biases in academe and government: "Thus, it is vastly easier to mount a study of the people or plants or hydrology or soils of a watershed than of their interactions." Consequently, it is rare to encounter integration of physical, biological, social, and health sciences in any phase of climate studies and services, from basic research to observations to interaction with users. *Our Common Journey* calls for putting more funds into the hands of place-based institutions with a mission of promoting policy-driven knowledge and know-how. In moving forward for climate-related services, attention should be paid to developing the infrastructure and staffing needed to develop and synthesize climate products based on diverse data sets. This recommendation follows directly from the guiding principle for climate services that if a climate service function is to improve and succeed, it should be supported by active research. It also requires support for education and training, for research on developing applications across disciplines, for identifying what products to produce, and for data sharing and serving and modeling infrastructure. At the same time, the ability and appetite to use climate products that provide syntheses across disciplines should be built by outreach to and education of users. At present, proposals for multidisciplinary climate research are difficult to fund in that they fall across too many different agencies. It is recommended that a mechanism for soliciting, reviewing, and funding such studies be put in place as soon as possible. Such a mechanism could be modeled after the National Oceanographic Partnership Program that is beginning to show progress across the public-private-academic interfaces.

Recommendation 3.3: Foster climate policy education Climate science has become increasingly interdisciplinary, involving meteorology, oceanography, terrestrial physics, and biology. Consequently, climate education should be more interdisciplinary. Climate services inherently involve interfacing climate science with user communities in such areas as agriculture, energy, public health, and government. The evolution of climate services will provide an increased demand for people who have training in both climate science and the social sciences. Universities should therefore initiate majors and minors in climate policy to enable informed planning and management of climate services. Such programs should include education in the basics of climate science, identifying the needs of various user communities and creating and disseminating useful climate information for them. This recommendation follows the education and outreach mission of universities under the guiding principles for climate services (see Chapter 3).

Recommendation 3.4: Enhance the understanding of climate through public education. Climate is increasingly important in the decisions made by individuals, corporations, cities, states, and the nation as a whole. Critical to the successful application of climate information is an educated consumer. The basis of predictions, understanding of uncertainties, and an ability to communicate probabilities are key elements in enhancing the benefit of climate information. Outreach and education are key elements of developing a strong climate service in the United States. *The Science of Regional and Global Change* (NRC 2001b) states that fundamental change will be possible only if education and outreach efforts communicate the progress of understanding: "The quality, diversity, inclusiveness, and timeliness of education and outreach efforts are probably the most important factors determining success or failure in the long run." BASC views this recommendation as directly responsive to the guiding principle for climate services: "Greater education of the user in the meaning and significance of climate information is likely to promote greater use and more robust application of that information" (NRC 2001b).

CONCLUDING REMARKS

BASC did not explicitly explore a formal climate services organizational structure within a specific federal, state, or local agency. Several proposals internal to the government have been made in the past. For example, the *NOAA Climate Services Plan* (Changnon et al. 1990) offers several suggestions for consolidating NOAA climate organizations, the Climate Prediction Center and the National Climatic Data Center, with the RCCs to form a unified climate service. However, an evaluation of such plans is outside the scope of this report. Instead, BASC reviewed current climate services and their potential evolution. The existing network of state climatologists, RCCs, national agencies, and private sector organizations has provided services in the past and provides increasingly competent services today. BASC believes that the principles discussed in this report represent the best practices of the various activities and that if applied across all levels of services—local, state, regional, and national—would improve the overall climate services to the nation. The recommendations contained in this report offer concrete first steps to a better integrated national system.

REFERENCES

Angel, J. 2000. President of the American Association of State Climatologists. Presentation to the September 7-8, 2000 meeting of the Climate Research Committee, National Research Council, Washington, D.C.

Berz, G. A. 1993. Global warming and the insurance industry. *Interdiscipl. Scien. Rev.* 18:120-125.

Changnon, S. A. 1995. Applied climatology: a glorious past–an uncertain future. *Preprints, 9th Conference on Applied Climatology*. American Meteorological Society, pp. 1-6.

Changnon, S. A. 1999. Impacts of 1997–98 El Niño-generated weather in the United States. *Bull. Amer. Meteor. Soc.* 80:1819–1827.

Changnon, S. A. (ed). 2000. El Niño 1997–1998: the Climate Event of the Century, New York: Oxford University Press.

Changnon, S. A., F. T. Quinlan, and E. M. Rasmusson 1990. NOAA Climate Services Plan. Climate Analysis Center, National Weather Service, Silver Spring, MD.

Crawford, K. 2000. The Oklahoma Climate Survey. Presentation to the August 7–11, 2000 meeting of the Board on Atmospheric Sciences and Climate, Woods Hole, MA.

Cuff, T. 2000. Navy Climate Service Programs. Presentation to the August 7–11, 2000 meeting of the Board on Atmospheric Sciences and Climate, Woods Hole, MA.

Dlugolecki, A. F. 1992. Insurance implications of climatic change. *The Geneva Papers on Risk and Insurance* 17:393–405.

Hecht, A. D. 1984. Meeting the challenge of climate service in the 1980s. *Bull. Amer. Meteor. Soc.* 65:365-366.

REFERENCES

Hooke, W., and R. A. Pielke, Jr. 2000. Short-term weather prediction: an orchestra in search of a conductor. In *Prediction: Science, Decision Making and the Future of Nature*, eds. D. Sarewitz, R. A. Pielke, Jr., and R. Byerly, 61–83. Washington, D.C.: Island Press.

HUD (U.S. Department of Housing and Urban Development). 1993. Frost-protected shallow foundations in residential construction, Phase 1. Instrument No. DU100K000005897 (April). Upper Marlboro, Maryland: NAHB Research Center.

IPCC (Intergovernmental Panel on Climate Change). 2001. Climate Change 2001: The Scientific Basis. Cambridge, U.K.: Cambridge University Press.

Insurance Research Council. 1995. Coastal exposure and community protection: Hurricane Andrew's legacy. Insurance Research Council, Inc., and Insurance Institute for Property Loss Reduction, Wheaton, IL, 48 pp.

Karl, T. R. 2000. Climate Services at the National Climatic Data Center. Presentation to the August 7–11, 2000 meeting of the Board on Atmospheric Sciences and Climate, Woods Hole, MA.

Karl, T. R., V. E. Derr, D. R. Easterling, C. K. Folland, D. J. Hoffman, S. Levitus, N. Nicholls, D. E. Parker, and G. W. Withee. 1995. Critical issues for long-term climate monitoring. *Clim. Change* 31:185-221.

Landsea, C. W., and J. A. Knaff. 2000. How much skill was there in forecasting the very strong 1997–98 El Niño? *Bull. Amer. Meteor. Soc.* 81:2107-2119.

Munich Re. 1994. Weather patterns: warming to disaster. Technical report, Munich Reinsurance, London.

NRC (National Research Council). 1992. *Toward a New National Weather Service—Second Report.* Washington, D.C.: National Academy Press.

NRC. 1995. *Natural Climate Variability on Decade-to-Century Time Scales.* Washington, D.C.: National Academy Press.

NRC. 1996. *Learning to Predict Climate Variations Associated with El Niño and the Southern Oscillation, Accomplishments and Legacies of the TOGA Program.* Washington, D.C.: National Academy Press.

NRC. 1998a. *The Atmospheric Sciences Entering the Twenty-First Century.* Washington, D.C.: National Academy Press.

NRC. 1998b. *Capacity of U.S. Climate Modeling to Support Climate Change Assessment Activities.* Washington, D.C.: National Academy Press.

NRC. 1998c. *Decade-to-Century-Scale Climate Variability and Change: A Science Strategy.* Washington, D.C.: National Academy Press.

NRC. 1998d. *Global Environmental Change: Research Pathways for the Next Decade.* Washington, D.C.: National Academy Press.

NRC. 1999a. *Making Climate Forecasts Matter.* Washington, D.C.: National Academy Press.

NRC. 1999b. *Adequacy of Climate Observing Systems.* Washington, D.C.: National Academy Press.

NRC. 1999c. *Our Common Journey: a transition toward sustainability.* Washington, D.C.: National Academy Press.

NRC. 1999d. *A Vision for the National Weather Service: Road Map for the Future.* Washington, D.C.: National Academy Press.

NRC. 2000a. *From Research to Operations in Weather Satellites and Numerical Weather Prediction: Crossing the Valley of Death.* Washington, D.C.: National Academy Press.

NRC. 2000b. *Issues in the Integration of Research and Operational Satellite Systems for Climate Research: Part I. Science and Design.* Washington, D.C.: National Academy Press.

NRC. 2000c. *Issues in the Integration of Research and Operational Satellite Systems for Climate Research: Part II. Implementation.* Washington, D.C.: National Academy Press.

NRC. 2001a. *Improving the Effectiveness of U.S. Climate Modeling.* Washington, D.C.: National Academy Press.

NRC. 2001b. *The Science of Regional and Global Change: Putting Knowledge to Work.* Washington, D.C.: National Academy Press.

NRC. 2001c. *Grand Challenges in Environmental Sciences.* Washington, D.C.: National Academy Press.

NWS (National Weather Service). 1991. Federal Register Notice: Policy Statement on Weather Service/Private Sector Roles. Federal Register Vol. 56, #13, p. 1984. Available at: <http://www.nws.noaa.gov/im/fedreg.htm>.

Pielke, Jr., R. A., and R. A. Pielke, Sr. 1997. *Hurricanes: their nature and impacts on society.* Chichester, England: John Wiley and Sons.

PricewaterhouseCoopers. 2001. The Weather Risk Management Industry: Survey findings for November, 1977 to March, 2001. A Report to the Weather Risk Management Association. Available at <http://www.wrma.org>.

Redmond, K. 2000. Activities of the Western Regional Climate Center. Presentation to the August 7–11, 2000 meeting of the Board on Atmospheric Sciences and Climate, Woods Hole, MA.

Rhodes, S. L., E. Ely, and J. A. Dracup. 1984. Climate and the Colorado River: the limits of management. *Bull Amer. Meteor. Soc.* 65:682-691.

Stiglitz, J, J. Orszag, and P. Orszag. 2000. The Role of Government in a Digital Age. Computer and Communications Industry Association, Washington, D.C. Available at: <http://www.ccianet.org/digitalgovstudy/main.htm>.

WeatherRisk. 2000. Supplement to *Risk* and *Energy and Power Risk Management*, August. New York: Risk Waters Group.

USGCRP (United States Global Change Research Program). 2000. *Our Changing Planet: The FY 2001 U.S. Global Change Research Program.* Washington, D.C.: National Science and Technology Council.

USGCRP. 2001. *Climate Change Impacts on the United States.* Cambridge, U.K.: Cambridge University Press.

ACRONYMS AND ABBREVIATIONS

BASC	Board on Atmospheric Sciences and Climate
CCM	Certified Consulting Meteorologist
CPC	Climate Prediction Center
DOE	(U.S.) Department of Energy
DOI	(U.S.) Department of the Interior
ENSO	El Niño/Southern Oscillation
FAA	Federal Aviation Administration
FCMSSR	Federal Committee for Meteorological Services and Supporting Research
FEMA	Federal Emergency Management Agency
HUD	(U.S.) Department of Housing and Urban Development
IPCC	Intergovernmental Panel on Climate Change
IRI	International Research Institute
NASA	National Aeronautics and Space Administration
NCDC	National Climatic Data Center
NCEP	National Centers for Environmental Prediction
NOAA	National Oceanic and Atmospheric Administration
NRC	National Research Council
NWS	National Weather Service
OCS	Oklahoma Climatological Survey
OFCM	Office of the Federal Coordinator for Meteorology
RCC	regional climate center

USDA   U.S. Department of Agriculture
USGS   U.S. Geological Survey
USGCRP   U.S. Global Change Research Program

BOARD MEMBERS' BIOGRAPHIES

ERIC J. BARRON (*Chair*) is Director of the EMS Environment Institute and Distinguished Professor of Geosciences at Pennsylvania State University. He received his Ph.D. in geophysics from the University of Miami. His professional experience encompasses fellow and scientist at the National Center for Atmospheric Research, associate professor of marine geology and geophysics at the University of Miami, and director of the Earth System Science Center at Penn State. His specialty is paleoclimatology/paleoceanography. His research emphasizes global change, specifically numerical models of the climate system and the study of climate change throughout Earth history. Dr. Barron is a fellow of the American Geophysical Union and the American Meteorological Society.

SUSAN K. AVERY is Director of the Cooperative Institute for Research in Environmental Sciences and Professor of Electrical and Computer Engineering at the University of Colorado, Boulder. She received her Ph.D. in atmospheric science from the University of Illinois. She has served on the faculty in the Department of Electrical Engineering, University of Illinois, and as associate dean of research and graduate education, College of Engineering, University of Colorado. Her specialty is atmospheric dynamics; her fields of research are wave dynamics, including the coupling of atmospheric regions and interactions between scales of motion, precipitation studies using ground-based radar; and the use of ground-based Doppler radar techniques for observing the neutral atmosphere. Dr. Avery is a fellow of the American Meteorological Society and the Institute of Electrical and Electronics Engineers, as well as a member of the American Geophysical Union. She is the past chair of the United States Committee to the International Union of Radio Science and a past officer of the University Corporation for Atmospheric Research.

RAYMOND J. BAN is Senior Vice President of Meteorological Affairs and Operations at The Weather Channel, Inc. (TWC). His responsibilities include oversight of the meteorological operations and all meteorological activities of the company. Prior to joining TWC in 1982, he was employed by Accu-Weather, Inc.. He graduated from Pennsylvania State University with a degree in meteorology. Mr. Ban is currently serving as Commissioner of Professional Affairs of the American Meteorological Society, as a member of the Science Advisory Committee of the U.S. Weather Research Program, and also serves on the Board of the College of Earth and Mineral Sciences at Penn State. He is a past member of the COMET (Cooperative Program for Operational Meteorology, Education and Training) Advisory Panel and the Research and Technical Committee of the Southeast Region Climate Center.

HOWARD B. BLUESTEIN is Professor of Meteorology at the University of Oklahoma, where he has served since 1976. He received his Ph.D. in meteorology from the Massachusetts Institute of Technology. His research interests are the observation and physical understanding of weather phenomena on convective, mesoscale, and synoptic scales. Dr. Bluestein is a fellow of the American Meteorological Society (AMS) and the Cooperative Institute for Mesoscale Meteorological Studies. He is past chair of the National Science Foundation Observing Facilities Advisory Panel, the AMS Committee on Severe Local Storms, and the University Corporation for Atmospheric Research Scientific Program Evaluation Committee, and a past member of the AMS Board of Meteorological and Oceanographic Education in Universities. He is also the author of a textbook on synoptic-dynamic meteorology and *Tornado Alley*, a book for the scientific layperson on severe thunderstorms and tornadoes.

STEVEN F. CLIFFORD is director of the NOAA Environmental Technology Laboratory. He received his Ph.D. in engineering science from Dartmouth College. One of his research goals is to develop a global observing system using ground-based, airborne, and satellite remote sensing systems to better observe and monitor the global environment and use these observations as input to global air-sea circulation models for improving forecasts of weather and climate change. Dr. Clifford is a member of the National Academy of Engineering. He is also a fellow of the Optical and Acoustical Societies of America, a senior member of the Institute of Electrical and Electronics Engineers, and a member of the American Physical Society, the American Geophysical Union, and the American Meteorological Society. He was the recipient of the 1998 Meritorious Presidential Rank Award.

GEORGE L. FREDERICK is General Manager of Vaisala Meteorological Systems Inc. in Boulder, Colorado. He received his M.S. in meteorology from the University of Wisconsin, Madison. Mr. Frederick manages a strategic business unit of Vaisala involved with atmospheric projects that include design, installation, and data processing of atmospheric measurement systems employing both in-situ and remote sensing techniques. He is working with government, state, and private industry to better employ remote sensing technology for the enhanced monitoring of atmospheric pollutants, aviation safety, and mesoscale weather forecasting. Mr. Frederick is a fellow and past president (1999–2000) of the American Meteorological Society.

MARVIN A. GELLER is a Professor of Atmospheric Science and the Dean and Director of Stony Brook's Marine Sciences Research Center (State University of New York). He received his Ph.D. from the Massachusetts Institute of Technology. Dr. Geller is a well-known researcher in atmospheric dynamics and serves on several national and international committees. He is co-chair of the World Climate Research Programme's Stratospheric Processes and Their Role in Climate (SPARC) project, president of the American Geophysical Union's Atmospheric Sciences section, co-chair of the National Association of State Universities and Land-Grant Colleges' Board on Oceans and Atmosphere, president of the International Council for Science Scientific Committee on Solar-Terrestrial Physics (SCOSTEP), and a fellow of the American Meteorological Society. He has also served on and chaired several National Research Council panels and committees, including the Committee on Solar-Terrestrial Research.

CHARLES E. KOLB is President and Chief Executive Officer of Aerodyne Research, Inc. He received his Ph.D. in physical chemistry from Princeton University and joined Aerodyne as a senior research scientist in 1971. His research interests have included atmospheric and environmental chemistry, combustion chemistry, materials science, and the chemistry and physics of rocket and aircraft exhaust plumes. Dr. Kolb is a fellow of the American Physical Society, the Optical Society of America, and the American Geophysical Union, where he served as the atmospheric sciences editor of *Geophysics Research Letters* (1996–1999). He is also a member of the American Association for the Advancement of Science and the American Chemical Society, which honored him with its Award for Creative Advances in Environmental Science and Technology in 1997. He has served on a number of National Research Council committees, including the Committee on Atmospheric Chemistry.

JUDITH L. LEAN is a research physicist at the Naval Research Laboratory. She received her Ph.D. in Atmospheric Physics from the University of Adelaide, Australia. She specializes in the study of the variability of solar radiation. The focus of her current research is the mechanisms, models, and measurements of variation in the sun's radiative output, and the effects of this variability on the earth's global climate and space weather. Dr. Lean is a member of the American Geophysical Union and the American Meteorological Society.

MARGARET A. LEMONE is a senior scientist at the National Center for Atmospheric Research (NCAR). She has two primary scientific interests: the structure and dynamics of the atmosphere's planetary boundary layer and its interaction with the underlying surface and clouds overhead, and the interaction of mesoscale convective with the boundary layer and surface underneath, and with the surrounding atmosphere. Dr. LeMone is a Fellow of the American Association for the Advancement of Science and the American Meteorological Society. She is also a member of the National Academy of Engineering (NAE) and has served on the National Research Council's Panel on Improving the Effectiveness of U.S. Climate Modeling and the Special Fields and Interdisciplinary Engineering Peer Committee of the NAE. Dr. LeMone received her Ph.D. in atmospheric sciences from the University of Washington.

MARIO J. MOLINA is a professor at the Massachusetts Institute of Technology. His interests are common to the fields of atmospheric chemistry, chemical kinetics, and photochemistry. His research is directed at understanding the potential implications of changes in the chemical composition of the earth's atmosphere and aims to elucidate the role of aerosols and clouds in the changing chemistry of the global atmosphere. In 1995 he shared the Nobel Prize in Chemistry with two co-investigators for their work in demonstrating the link between man-made chlorofluorocarbons in the atmosphere and the damage to the ozone layer. Dr. Molina is a member of the National Academy of Sciences and the Institute of Medicine. He has served on the President's Committee of Advisors on Science and Technology, the Secretary of Energy Advisory Board, and on several National Research Council committees, including the Board on Environmental Studies and Toxicology and the Committee on Atmospheric Chemistry. Dr. Molina received his Ph.D. in physical chemistry from the University of California, Berkeley.

ROGER A. PIELKE, JR. is an Associate Professor in the Environmental Studies Program and a Fellow of the Cooperative Institute for Research in the Environmental Sciences (CIRES) at the University of Colorado, Boulder. He is overseeing the development of a new Center for Science and Technology Policy Research. Previously, Dr. Pielke was a scientist at the National Center for Atmospheric Research where he studied societal responses to extreme weather events, policy responses to climate change, and U.S. science policy. With a B.A. in mathematics and a Ph.D. in political science, he focuses his research on the relation of scientific information and public and private sector decision making. His current areas of interest include technology policy in the atmospheric and related sciences, use and value of prediction in decision making, and policy education for scientists. Dr. Pielke chairs the American Meteorological Society's Committee on Societal Impacts and serves on the Science Steering Committee of the World Meteorological Organization's World Weather Research Programme. He sits on the editorial boards of *Policy Sciences*, *Bulletin of the American Meteorological Society*, and *Natural Hazards Review*. He is a co-author or co-editor of three books, most recently *Prediction: Decision making and the future of nature*.

MICHAEL J. PRATHER is a professor in the Earth System Science Department at the University of California, Irvine. He received his Ph.D. in astronomy from Yale University. His research interests include the simulation of the physical, chemical, and biological processes that determine atmospheric composition, specifically ozone and other trace gases. Dr. Prather has authored chapters in the World Meteorological Organization's Ozone Assessments (1985–1994) and the IPCC's assessments of climate and aviation effects (1995–2000). He is a fellow of the American Geophysical Union and a foreign member of the Norwegian Academy, and has served on several National

Research Council committees, including the Panel on Climate Variability on Decade-to-Century Time Scales.

WILLIAM J. RANDEL is a senior scientist at the National Center for Atmospheric Research. His research interests include dynamic variability and climatology of the stratosphere and the observed variability of trace constituents in the middle atmosphere using satellite observations. He has contributed to the World Meteorological Organization/United Nations Environment Programme (WMO/UNEP) Assessments of ozone and temperature trends in the stratosphere and is actively involved with a number of Stratospheric Processes and Their Role in Climate (SPARC) activities. He has also served as chair of the American Geophysical Union's Committee on Atmospheric Dynamics and the American Meteorological Society's Committee on the Middle Atmosphere. Dr. Randel received his Ph.D. in physics from Iowa State University.

ROBERT T. RYAN is Chief Meteorologist at WRC-TV (NBC 4) in Washington, D.C. He received his Masters in atmospheric sciences from the State University of New York, Albany. Prior to his career in broadcasting, he was a research associate in the Physics Section at Arthur D. Little where his work involved various cloud physics projects with the Department of Defense. He was also involved in various meteorological field experiments for NASA and the U.S. Army. Mr. Ryan is nationally and internationally recognized for his outreach and educational activities in meteorology and atmospheric sciences. He was the first broadcast meteorologist elected to serve as president of the American Meteorological Society (AMS). He has also served the AMS as Councilor, Commissioner of Professional Affairs, and Chairman of the Board of Broadcast Meteorology. In 1997, Mr. Ryan received the Charles Franklin Brooks Award for service to the Society. He has also served on various AMS, NASA, and NOAA study groups, testified a number of times before Congress, and is the winner of nine Emmy awards for his television productions and service to the community. He is a fellow of the AMS and a member of the American Association for the Advancement of Science.

MARK R. SCHOEBERL is a senior atmospheric scientist at NASA's Goddard Space Flight Center. He received his Ph.D. in physics from the University of Illinois. He has also served as a scientist at Science Applications, Inc., and at the Naval Research Laboratory. His fields of research include atmospheric dynamics, stratospheric physics and chemistry, and numerical modeling. He is the EOS-Aura Project Scientist. Dr. Schoeberl has received many NASA awards, including the Group Achievement Award in 1988, 1989, 1991, 1994, 1995, and 1998, and the Education and Outreach Award in 1999. In addition he was awarded the Scientific Achievement Medal in 1991, the Distinguished Service Medal in 2000, and the William Nordberg Award for Earth Sciences in 1998. He is a fellow of the American Geophysical Union (AGU) and past president of the AGU's Atmospheric Sciences section. He is also a fellow of the American Association for the Advancement of Science and the American Meteorological Society.

JOANNE SIMPSON is Chief Scientist for Meteorology and Senior Fellow at NASA's Goddard Space Flight Center. She received her Ph.D. from the University of Chicago and an honorary D.Sc. from the State University of New York, Albany. She has served on the faculty at the University of California, Los Angeles, and at the University of Virginia. She was also head of the atmospheric physics and chemistry laboratory at ESSA, director of NOAA's Experimental Laboratory, and head of the severe storms branch at NASA Goddard. Her areas of research include atmospheric convection, tropical meteorology, weather modification, and satellite meteorology. Dr. Simpson is a member of the National Academy of Engineering and a fellow and honorary member of the American Meteorological Society (AMS). She has received many awards, including the Charles Franklin Brooks Award and the Meisinger Award from the AMS, the Rossby Research Medal, the Silver and Gold Medals from the Department of Commerce, and NASA's Exceptional Scientific Achievement Medal and William Nordberg Award. She has served on several National Research Council committees including the Committee on Climate, Ecosystems, Infectious Diseases, and Human Health.

THOMAS F. TASCIONE is Vice President, Weather Systems Operations, Sterling Software (U.S.), Inc. In this position he oversees the development of a state-of-the-art weather forecasting technology for the Defense Department. In addition, he manages a commercial space weather forecasting service to support the commercial satellite and electric power industries. Another focus area is advanced weather visualization technology for aviation including a patented system to extract and apply weather information along a route a flight using a simple web-browser. His prior professional experience was with the Department of Defense (1972–1993), during which he held numerous weather and space weather forecasting positions. He was the architect of the Air Force space weather forecast models program, and he co-chaired the interagency committee that initiated and developed the National Space Weather Program (NSWP). Dr. Tascione received his Ph.D. in space physics from Rice University.

ROBERT A. WELLER is a Senior Scientist at Woods Hole Oceanographic Institution, where he holds the Secretary of the Navy/CNO Chair in Oceanography and is Director of the Cooperative Institute for Climate and Ocean Research. He received his Ph.D. in Physical Oceanography from the University of California, San Diego, Scripps Institution of Oceanography. His research interests include wind-forced motion in the upper ocean; mixed layer dynamics; upper ocean velocity structure studies; air-sea interaction; the role of the ocean in climate; and the development of upper ocean and surface meteorological instrumentation and platforms for air/sea experiments. Dr. Weller received the James B. Macelwane Award from the American Geophysical Union (AGU) in 1986. He is a fellow of the AGU and president of the Ocean Sciences section. He is also a member of the American Association for the Advancement of Science, the American Meteorological Society, and the Oceanographic Society.

ERIC F. WOOD is Professor of Civil and Environmental Engineering at Princeton University, where he has taught since 1976. He received his Sc.D. in civil engineering from the Massachusetts Institute of Technology. His areas of research include hydroclimatology with an emphasis on land-atmospheric interaction, hydrologic remote sensing, and hydrologic impact of climate change. Dr. Wood is a member of the National Research Council's Climate Research Committee and Committee on Hydrologic Sciences. He is a member of the Council and a fellow of the American Meteorological Society (AMS) and a fellow of the American Geophysical Union (AGU). He has received the AGU Robert E. Horton Award, the AMS Horton Lectureship, and the Princeton Rheinstein Award.

Appendix A

STATEMENT OF TASK

The Board on Atmospheric Sciences and Climate will convene a workshop and subsequently produce a report that articulates the fundamental elements that are required in the delivery of climate services. The audiences for this study are those federal, academic, and private individuals and organizations that are currently providing or are capable of providing climate information on all time scales. This study will:

- Define climate services.
- Describe potential audiences and providers of climate services.
- Describe the types of products that should be provided through a climate service.
- Outline the roles of the public, private, and academic sectors in a climate service.
- Describe fundamental principles that should be followed in the provision of climate services.

The workshop will include a broad array of both providers and users of climate information. An informal discussion paper that outlines fundamental issues concerning the provision of climate services will be prepared in advance of the meeting.

Appendix B

WORKSHOP PARTICIPANTS

Susan Avery	University of Colorado, Boulder
Eric Barron	Pennsylvania State University
Howard Bluestein	University of Oklahoma
Kenneth Crawford	Oklahoma Climate Survey
Thomas Cuff	Office of the Oceanographer of the Navy
Elbert (Joe) Friday	National Research Council
Marvin Geller	State University of New York, Stony Brook
Diane Gustafson	National Research Council
Michael Hall	NOAA
James Hansen	NASA
Eugenia Kalnay	University of Maryland
Thomas Karl	NOAA
Jack Kelly	NOAA
Charles Kolb	Aerodyne Research, Inc.
William Lau	NASA
Peter Leavitt	WSC (retired)
Ants Leetmaa	NOAA
David Legler	U.S. CLIVAR
Stephen Lellyett	Bureau of Meteorology (Australia)
Robert Livezey	NOAA
Greg Mandt	NOAA
Eugene Rasmusson	University of Maryland

APPENDIX B

Kelly Redmond   *Western Regional Climate Center*
Michele Rienecker   *NASA*
Robert Ryan   *WRC-TV (NBC 4)*
Edward Sarachik   *University of Washington*
Robert Schiffer   *NASA*
Peter Schultz   *National Research Council*
Daniel Tarpley   *NOAA*
Elke Weber   *Columbia University*
Robert Weller   *Woods Hole Oceanographic Institution*
Gregory Withee   *NOAA*
Eric Wood   *Princeton University*

Appendix C

AGENDA

Note: The agenda as presented is general, with no explicit times allocated for any item. The subjects listed are those to be covered in the summer study, but the agenda is flexible to permit in depth pursuit of an issue as needed or less time to less difficult issues. All participants are encouraged to actively engage in the discussion items and the presentations.

TUESDAY, AUGUST 8, 2000

8:30 a.m.   Welcome and Introductions (Barron)

Logistics for the meeting (Friday and Gustafson)

Scope of the study (Barron)

Context of the national needs and priorities for climate services, including *The Atmospheric Sciences Entering the Twenty-First Century*, *Making Climate Forecasts Matter*, and other applicable NRC reports. This should include a brief discussion of the state of the climate services capability, the promises for the near future, and the potential for new service families. (Barron, Rasmusson)

NOAA climate services views:

APPENDIX C

- NWS perspective (Kelly, Mandt)
- NESDIS perspective (Withee)
- OAR perspective (Hall)
- A National Climate Service through the eyes of NCDC (Karl)

Navy perspective (Cuff)

NASA climate services views (Schiffer)

Making Climate Forecasts Matter (Weber)

Reanalysis, the view from CLIVAR (Eugenia Kalnay)

WEDNESDAY, AUGUST 9, 2000

8:30 a.m.   Perspectives of climate services (users, requirements, potentials, shortfalls):

- State view (Crawford)
- Regional view (Redmond)
- Private sector view (Leavitt)
- International issues (Lellyett)

Data as a service, the need for a 'data priesthood' (Discussion led by Gene Rasmusson)

Assessments as a service (Discussion led by Barron)

Roles and missions: public, private, and academic sectors (Discussion led by Frederick)

Examinations of future demands for climate services (Discussion led by Barron)

THURSDAY, AUGUST 10, 2000

8:30 a.m.   Open discussion during which the issues that have been presented during the previous days will be crystallized and a general report outline will be prepared and discussed. The conclusions and recommendations will not be formulated during this open session; rather, this will be an opportunity for clarification of the material presented and the BASC's understanding of the material and issues. Agency participants are urged to be actively involved during this session.

- Definition of climate services
- Identification of users and providers
- Types of products included in climate services
- Future demands for climate services
- Roles and missions, public, private, and academic roles and responsibilities

FRIDAY, AUGUST 11, 2000

8:30 a.m. Discussion continued

Fundamental principles to be followed in the provision of climate services:

- Observations
- Data sets
- Modeling
- Information and communication
- Assessments

Objectives and priorities

Structure, overlap with environmental services, etc.

SATURDAY, AUGUST 12, 2000 *CLOSED SESSION*

8:30 a.m. Report Preparation: Formulation of findings and recommendations.

1:00 p.m. Adjourn

Appendix D

EXAMPLES OF AREAS OF CLIMATE INFORMATION REQUESTS TO THE WESTERN REGIONAL CLIMATE CENTER

AGRICULTURE/LIFE SCIENCES (GROUP 1)

Plant disease
Cereals/corn/berries/grasses/
ornamentals/nuts/mints/melons/
fruits/vegetables/hay/alfalfa/
tubers/mushrooms/spices/

Plant Growth
Planting times
Germination
Dormancy requirements
Frost probabilities
Lodging
Harvest conditions

Product quality
Seed spoilage
Transport conditions
Storage conditions

Product marketing

Chemical tests
Pesticides
Growth retardants
Growth enhancers

Seed certification

Relocations
New crop introduction
Climate changes/fluctuations

Erosion
Water
Wind

Viticulture
Seed companies
Soil climatologies
Soil chemistry
Degree days - growing/chilling

Water issues
Consumptive water use
Water stress
Drought
   Frequency
   Assessment
   Designation
Groundwater recharge rate
Groundwater use rate
Groundwater contamination
Irrigation needs
Soil water balance
Evapotranspiration
Evaporation climatologies
Runoff and non-point pollution

Insects
Moths/worms/beetles/flies/ants/
mites/maggots/grasshoppers/
crickets/caterpillars
Chilling hours/dormancy
Egg-laying conditions
Growth/hatch timing
Dispersion
Migration
Host plant environment
Pollination conditions
Metabolism rates (= temperature)
Pesticide effectiveness
Introduction of new pests
   Helpful (deliberate)
   Harmful (accidental)

APPENDIX D

AGRICULTURE / LIFE SCIENCES (GROUP 2)

Wildlife
Severe winters/summers
Habitat conditions
Migration
Transplantation
Breeding success
Birthing/calving success
Endangered species conditions
Refuge management

Fish
Lethal/injurious water
temperatures
  In streams and rivers
  Behind impoundments
  Ice effects
Weather-induced sediment loading
Passage time – anadromous species
Flow volume and timing
Ocean conditions
Hatchery conditions
Disease outbreaks
Condition of redds/eggs

Grazing and forage conditions

Caged and penned animals
Permanent (domesticated species)
Temporary
(transplants/relocations)

Bird counts
Fungus distributions

Landscaping
Christmas trees
Riparian (stream) conditions

Experiment stations-research
General data bases
Conditions during experiments

Forestry
Reforestation
  Viability of nursery stock
  Clearcut/canopied microclimates
Ecosystem management
Parkland grazing conditions
Regeneration rates
Tree-ring growth and density
Fire
  Ignition and growth potential
  Triggering events
  Firefighting conditions
  Labor force
  Equipment deployment
  Mop-up, restoration, reseeding
  Erosion susceptibility
  Frequency assessment
  Insect kills
  Descriptive indices (e.g., Haines)
  Lightning climatologies
  Slash fire planning
Timber sale requirements
Blowdowns
Long-term climate variability

ENGINEERING (GROUP 1)

Energy
Audits
Heat loss calculations
Utility costs
Users
   Cities/counties/companies/
   private citizens
Providers
   Utilities
     Hydropower supply
     Rate setting
     Energy demand
     Fuel planning
     Strategic planning

Alternative energy (climate-sensitive sources)
Wind - means & extremes
Passive solar
Small head hydro
Heat pumps
Passive cooling
Rel Hum, alt. fuel motor

Construction
Scheduling
   Equipment inventories
   Personnel hiring
   Outdoor painting
   Environmental conditions

Product testing
Specific conditions needed
Specific conditions not needed

Fog instruments
Corrosion tests

Uniform Building Codes

Hazardous phenomena
Tornadoes
Lightning
Hail
Ice storms
Tropical storms

Depth of frozen soil

Balloon and helicopter logging
Hazard likelihood
Performance standards

Power line routing
Stress on long atmospheric tethers

Airports
Runway orientation
Runway length
Number of runways needed

Instrumentation questions
Diesel low-temperature additives
Chip manufacturers
Vinyl glue separation
Electric field studies

APPENDIX D

ENGINEERING (GROUP 2)

Design criteria
Roofs
Culverts, bridges, etc.
Storm sewers
Sanitary sewers
Aquatic center pools
City and industrial ponds
  Cooling ponds
  Settling ponds
  Sewage treatment
  Hazardous waste containment
  Mine tailings
  Evaporation calculations
Lighting
Freeze/thaw cycle climatologies
Frost effects
Drifting snow (depth/orientation)
Dam design

Boiler capacity
Refrigeration needs
Generators
Greenhouse heating/cooling
Structure orientation
Structure strength
Pollution dispersion
Freeze probabilities
Excessive values of
  Heat
  Cold
  Wind
  Rain
  Snowfall
  Snow depth
  Humidity
Wave Erosion – causeways

LEGAL

Accidents
Cars
Motorcycles/bicycles
Airplanes
Railroads
Hang gliders
Falls on ice

Storm damage
Claims adjustors
Real cause of damage??
"Act of God" or expected??
Crop damage
  Wind
  Hail
  Heavy rain
Ocean waves
  Open seas/beaches
Event insurance claims
  Outdoor gatherings/events

Environmental Impact Statements
Endangered Species Act needs
Biological Opinions
Grazing allotment decisions
Ecosystem Management background
Hazard Rankings
Environmental Assessments
Wetlands determination

Construction overruns
Landslides
Shipment delays/difficulties

Pesticide drift

Crime conditions
Murder/assault/violent crimes
Decomposition rates
Burglaries
Traffic tickets
Evidence reconstruction

Water
Landfill runoff
Landfill or waste seepage
Frozen/broken pipes
Subdivision runoff
Landlord - tenant disputes
  Leaky roofs
  Storage of household goods
Industrial painting disruptions
Dike/containment breaches
Seed spoilage

Highway sanding/plowing conditions
Pollutant transport
Firefighting/rescue conditions
Cement hardening conditions
Health/workman's comp. Claims

ECONOMIC DEVELOPMENT AND OTHER

Manufacturing/business development
Design criteria
Construction conditions
Marketing and sales impacts
Inventory deployment
Siting of shipping facilities

Relocations from afar
Businesses
Manufacturing plants

Retirement decisions

Weather-sensitive products
Marketing decisions

Agribusiness - development of:
New crops
New products
New markets

Outdoor gatherings
Festivals
Concerts
Air shows
Auto/air/water/foot races

Motion picture filming conditions

Hiring of labor
Seasonal industries
　Construction
　Agriculture/migrant laborers
　Forestry
　Recreation

News media
Magazines
Newspapers
Radio
TV
Trade publications

Historical event conditions

Tourism
Vacation planning
Recreation climatology
　Hiking/camping/backpacking/
　rafting/bicycling/skiing/wind-
　surfing/fishing/hunting/
　mountain climbing/boating

Health
Relocation influences
　Skin problems
　Asthma/respiratory
　Allergies
　Trace chemical sensitivity
　Solar exposure
　　Melanomas
　　UV effects on vision
　　Cloudiness climatologies
　　Altitudinal variation of
　　radiation

Chambers of Commerce

Report Inclusions

National Weather Service
Background Information
Local forecasting studies/tools

Classroom/Educational
Other states/countries
Local climatologies

Climate trends
Yearly/decadal fluctuations
Regional climates
El Niño /Southern Oscillation
Global climate change

Home energy and gardening needs

Brochures

Interpretative public displays

General advice and interpretation